Computer Numerical Control

Computer Numerical Control

Herman W. Pollack
Professor Emeritus
Orange County Community College
Middletown, N.Y.

Terrance Robinson
Program Instructor

PRENTICE-HALL, Englewood Cliffs, N.J. 07632

Library of Congress Cataloging-in-Publication Data

Pollack, Herman W.
    Computer numerical control / Herman W. Pollack : Terrance
Robinson, program instructor.
      p. cm.
    ISBN 0-13-168378-0
    1. Machine-tools—Numerical control. I. Robinson, Terrance.
II. Title.
TJ1189.P64 1990
621.9′023—dc20   89-28151
                                                    CIP

Editorial/production supervision and
  interior design: Eileen M. O'Sullivan
Cover design: Wanda Lubelska
Manufacturing buyer: Gina Chirco Brennan

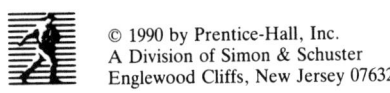

© 1990 by Prentice-Hall, Inc.
A Division of Simon & Schuster
Englewood Cliffs, New Jersey 07632

All rights reserved. No part of this book may be
reproduced, in any form or by any means,
without permission in writing from the publisher.

Printed in the United States of America

10 9 8 7 6 5 4 3 2 1

ISBN 0-13-168378-0

PRENTICE-HALL INTERNATIONAL (UK) LIMITED, *London*
PRENTICE-HALL OF AUSTRALIA PTY. LIMITED, *Sydney*
PRENTICE-HALL CANADA INC., *Toronto*
PRENTICE-HALL HISPANOAMERICANA, S.A., *Mexico*
PRENTICE-HALL OF INDIA PRIVATE LIMITED, *New Delhi*
PRENTICE-HALL OF JAPAN, INC., *Tokyo*
SIMON & SCHUSTER ASIA PTE. LTD., *Singapore*
EDITORA PRENTICE-HALL DO BRASIL, LTDA., *Rio de Janeiro*

Contents

	PREFACE		**xi**
CHAPTER 1	**COMPUTER NUMERICAL CONTROL**		**1**
	1.1	A History 1	
	1.2	Numerical Control 3	
	1.3	Computer Numerical Control 7	
	1.4	Codes 8	
	1.5	Direct Numerical Control 9	
	1.6	CAD/CAM and CIFM 10	
	1.7	Automatic Adaptive Control 12	
		Questions and Problems 13	
CHAPTER 2	**COMPUTER MATHEMATICS**		**15**
	2.1	Circle and Straight Line 15	
	2.2	Right Triangle 16	
	2.3	Special Relationships 18	
	2.4	Law of Sines 21	

2.5	Law of Cosines 22	
2.6	Compensation Applications: The Milling Machine 23	
	Questions and Problems 28	

CHAPTER 3   TIME TO MACHINE: FORCES   30

3.1	Cutting Speeds 30
3.2	Speed and Feed Tables 31
3.3	Feed Rate and Time to Machine 32
3.4	Horsepower Requirements 37
	Questions and Problems 41

CHAPTER 4   CUTTING TOOLS: THEIR USES   44

4.1	Theory of Cutting 44
4.2	Tool Bit Material 48
4.3	Terminology, Clearance, Rakes, and Chip Breakers 51
4.4	Milling Cutters 57
4.5	Drills and Reamers 60
	Questions and Problems 61

CHAPTER 5   CONTROL CENTERS   64

5.1	The Control Panel 64
5.2	Manual Data Input 64
5.3	Manual Control 66
5.4	The Floating Zero 66
5.5	Tool Length Offset 67
5.6	Feed Hold, Cycle Start, AND Single Block 69
5.7	Emergency Stop 69
	Questions and Problems 69

CHAPTER 6   ADDRESSES, CODES, BLOCKS, AND LINES   72

6.1	Definitions 72
6.2	Line 73

Contents   vii

      6.3   G Codes   74
      6.4   X, Y, and Z Codes   75
      6.5   I, J, and K Codes   77
      6.6   S Code   78
      6.7   F Code   78
      6.8   T Code   79
      6.9   M Code   80
      6.10   D And H Codes   80
      Questions and Problems   81

CHAPTER 7   INCREMENTAL CNC   83

      7.1   Command Systems   83
      7.2   Codes   88
      7.3   Program   92
      7.4   Cutter Position: X And Y Incremental Movements   94
      7.5   Linear Contouring   95
      7.6   Z Movement   99
      Questions and Problems   109

CHAPTER 8   CIRCULAR CONTOURING: 116
INCREMENTAL MODE

      8.1   Circular Contouring   116
      Questions and Problems   127

CHAPTER 9   ABSOLUTE MODE: MILLING MACHINE   132

      9.1   Absolute System   132
      Questions and Problems   143

CHAPTER 10   LINEAR INTERPOLATION: DRILL ROUTINES   145

      10.1   SPOT Drill   145
      Questions and Problems   152

CHAPTER 11 TOOL POSITIONING AND THREADING — 157

11.1 Tool Positioning And Tool Length Offset 157
11.2 Tool Positioning: Tapping 158
11.3 Programmable Cycle Files 162
Questions and Problems 168

CHAPTER 12 CIRCULAR INTERPOLATION: ABSOLUTE MODE — 173

12.1 Circular Interpolation 173
12.2 Multiquadrant Circular Interpolation 180
Questions and Problems 191

CHAPTER 13 RADIUS COMPENSATION: MILLING MACHINE — 194

13.1 Radius Compensation 194
13.2 Ninety-Degree Movements 194
13.3 Angular Compensation 200
13.4 Circular Interpolation 207
Questions and Problems 212

CHAPTER 14 RADIUS–ANGLE COMBINATIONS — 215

14.1 Radius–Angle Compensation 215
14.2 Radius–Angle–Radius Compensation 221
Questions and Problems 227

CHAPTER 15 CANNED CYCLES — 231

15.1 Cycles 231
15.2 Surface Milling 231
15.3 Canned Cycle: Multiple Row Drilling 235
15.4 Canned Cycle: Circular Pocket Milling 237
15.5 Canned Cycles: Pocket Milling 241
15.6 Polar Coordinates 245
Questions and Problems 248

CHAPTER 16   THE LATHE: RADIUS COMPENSATION   252

    16.1   Linear Contouring   252

    16.2   Circular Interpolation   260

    16.3   Radius–Angle Combinations   265

           Questions and Problems   275

GLOSSARY   280

INDEX   285

Preface

This text has been written because of a need for an applied approach to NC programming. It appeared to us that there was a need for a text that dealt with programming as a tool that could be applied directly to the computer operation of machine tools. It is true that every company has its own codes for activating specific functions. It is also true that the underlying programming methods are very similar. Therefore, it should be possible to develop text materials that are easily transferable to other systems. It is with this in mind that this text was written.

In essence, this text deals with broad categories such as: the basic items and tools needed to develop programs for CNC; the techniques of incremental and absolute programming; the program for the milling machine and the lathe; shortcuts such as canned cycle.

The authors wish to acknowledge the valuable assistance of the people at Prentice Hall Inc. for their efforts on behalf of this project. Special thanks go to Eileen O'Sullivan. Her dedication to excellence is to be commended. From H.W.P. a special thanks to Bette Holmes and Edith Luft, both of whom understood my moods, when it was necessary for me to devote blocks of time to this project. From Terry to Susan, a special thanks for her understanding when I took time away from her in order to work on the manuscript.

Herman W. Pollack

Terrance Robinson

Computer Numerical Control

1

Computer Numerical Control

1.1 HISTORY

The need to automate machinery developed a demand for, among other things, high production, greater precision, and greater competition in the marketplace evolved. These needs have pushed us to search for ways in which production depended more on machines than on human capabilities.

So it was that the need for machining round cross sections led to the development of the early lathe shown in Fig. 1.1. Early in the eighteenth century, John Wilkenson built a machine for boring holes, and Jacquard built and incorporated a mechanism for the loom that could be used to change, or control, patterns on materials being woven into cloth. He used a type of punch card.

In the middle of the eighteenth century, John Parsons built an automatic piano. It used a wide roll of paper with prepunched holes. As a bellows pumped air through the holes, the piano keys were activated, causing a melody to be played. Today, holes in a paper tape allow light to pass through, thus activating switches or servomotors. These, in turn, control the feeds and speeds of machine components, or tools, to perform dictated operations.

Much later, cutting tools were controlled by cams and hydraulic tracers. Coupled with this came a demand for greater production and better precision. Soon thereafter, assembly line automation, such as took place with the advent of the automobile, was developed to service

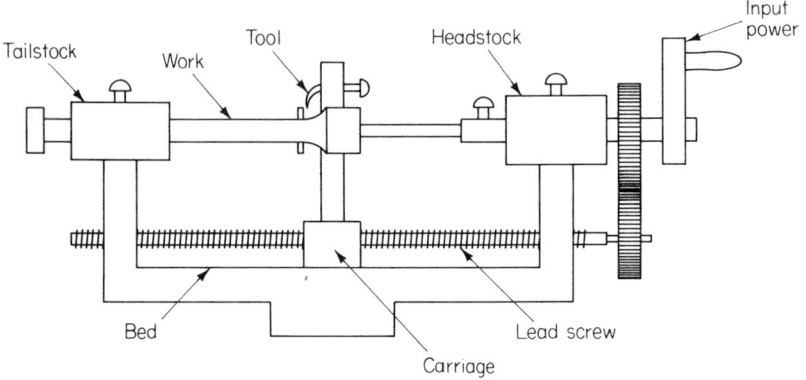

Figure 1.1

this need. A larger percentage of the human effort and decision making was transferred to the machine tool and later to the "non-thinking" control mechanisms. Production and precision increased.

In the early 1940s, production requirements became more stringent because of the demands of the U.S. Air Force. Aircraft were more complicated and sophisticated. Not only were production requirements greater, but the precision needed to machine more complicated shapes exceeded industry's capability to produce them. To meet these needs, the U.S. Air Force commissioned John Parsons to build a machine that could meet these demands. In 1950, John Parsons, working in cooperation with the Massachusetts Institute of Technology, produced a control system and a single-axis machine tool. Two years later they produced a three-axis machine tool. The latter used a prepunched tape. These machines, now called numerical control (NC), used complicated control interfaces. The control consisted of a combination of tubes and mechanical devices.

Once started, the refinements came quickly. Vacuum tubes were replaced with electronic tunes, then with solid-state circuitry. Refinement of the circuitry was also swift. Storage systems were refined and the concept of "read-only memory" (ROM) was developed. The ability to store materials in the memory of the computer led to computer numerical control (CNC). The ability to store blocks of commands in memory (canned cycles) and to retrieve these command sequences when needed made possible easy editing and retrieval, and reduced punch-tape reading errors.

The development of the microprocessor and the microcomputer helped to reduce the size of the control system, increase its memory and reliability, and greatly increase storage capacity. Editing and monitoring could now be done with visual displays. Built-in alarm systems, such as crash barriers that prevented cutting tools from collision, were now possible. Feeds and speeds, tool positions, and tool movements could now be changed and controlled.

At the start, linear and circular interpolation were accomplished using the conventional X, Y, and Z axes and Cartesian coordinates. Later, polar coordinates were used to define tool movements. Presently, direct numerical control (DNC), voice numerical control, and bubble memory are in the developmental stages.

Direct numerical control wires a computer directly into the machine tool, or a series of machine tools. With voice numerical control the operator reads the program into the computer, which then loads it into the memory bank in the machine. In both cases the machine is ready for operation.

Bubble memory makes use of a garnet crystal to store information. Much more material can be stored than is possible with other methods. Also, faster access to this material is provided. A greater number of canned cycles can be stored, making possible greater and more refined mathematical computations. In turn, better and more accurate cutter and tool nose radius compensation movements, in-process inspection, and compensation for error are possible.

1.2 NUMERICAL CONTROL

Historically, machine tools were controlled with handles, or levers, by an individual. An electric motor supplied the power to move a workpiece either linearly or in rotary fashion. The power was also directed to a mechanism that rotated and/or moved the workpiece. Thus, on the lathe [Fig. 1.2(a)], the workpiece is rotated as the cutting tool is caused to move linearly into the work. The milling machine [Fig. 1.2(b)], has the work mounted on the table. As the table moves linearly toward the cutter, upon contact with the work, the work is machined. Dimensions such as length, width, and depth are controlled by the operator.

In the late nineteenth century and the early twentieth century, automation was in the form of production milling machines that used cams and preset stops to control the motion of the table. Production lathes were characterized by capstans. Other forms of production machines were turret lathes [Fig. 1.2(c)], copy machines, tracers, and so on. The controls were mechanical, electrical, hydraulic, or a combination of these. They provided the basis for mass production, interchangeability, repeated dimensional accuracy, higher production rates, reduced labor and labor costs, and so on.

The numerically controlled (NC) machine tool (Fig. 1.3) may be

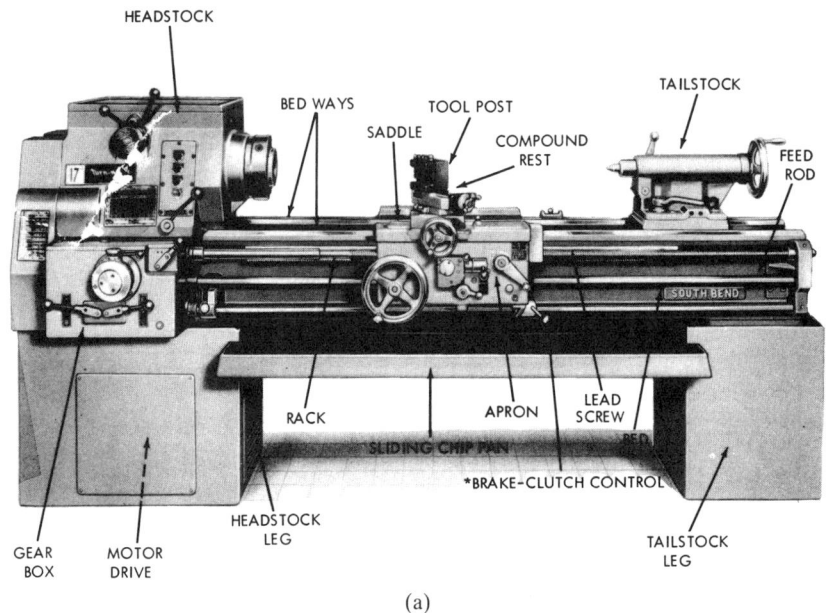

Figure 1.2(a) (*Photo courtesy of South Bend Lathe, Inc.*)

Figure 1.2(b) (*Courtesy of Cincinnati Milacron, Inc.*)

(c)

Figure 1.2(c) (*Photo courtesy of Clausing Company.*)

open-loop, closed-loop, or a combination of open- and closed-loop. In the open-loop machine [Fig. 1.4(a)], a command is sent to the tool, or table, that commands it to move a certain distance rapidly, or slowly, and stop. The movement is made. The tool, or table, waits for the next command.

Since the input signal is in an open-loop system, there is no feedback signal. Therefore, there is no way the system can determine whether or not the signal was executed as intended. The system cannot verify whether the command did, or did not, commit an error. The turret lathe is an example of an open-loop system. Cams and stops may control the motion of the turret, but there is no way that the system can tell whether, or not, the tool has performed the desired operation. To make the determination, the machinist must measure the workpiece.

The closed-loop system [Fig. 1.4(b)] has a feedback line from a sensor. When the table has reached its desired position, the sensor will detect this and direct the table to stop.

NC machines are controlled by a prepunched paper tape or through manual input. The paper tape uses letters, numbers, or symbols to tell the machine what is to be done. The closing, or opening, of a circuit is related to the presence, or absence, of a hole in the tape.

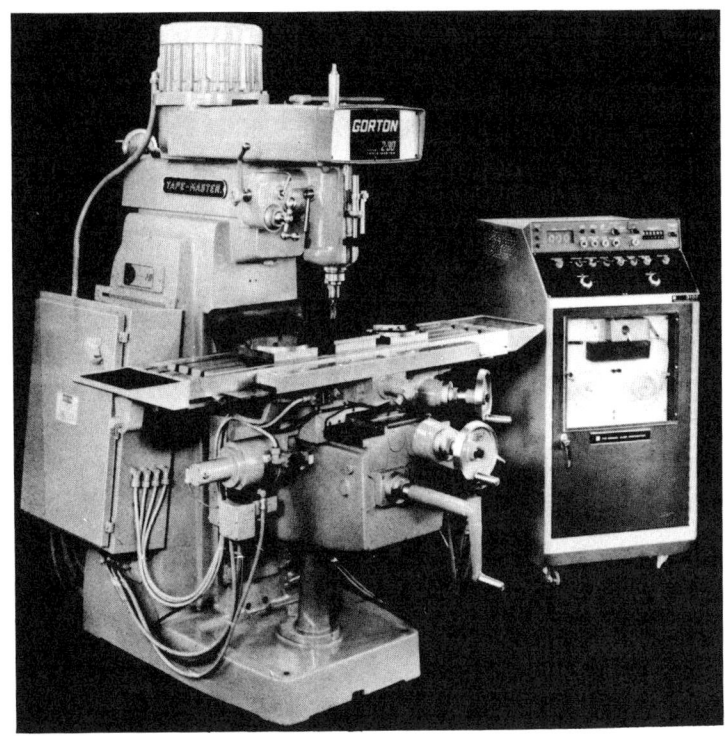

Figure 1.3 (*Courtesy of George Gorton Machine Company.*)

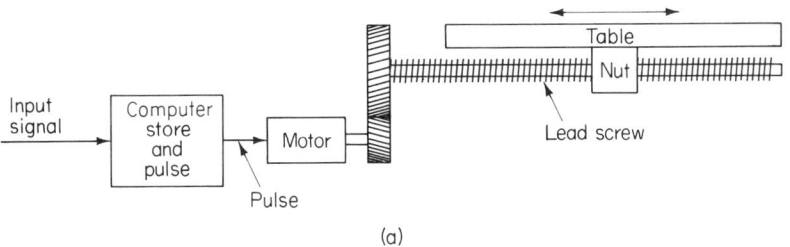

(a)

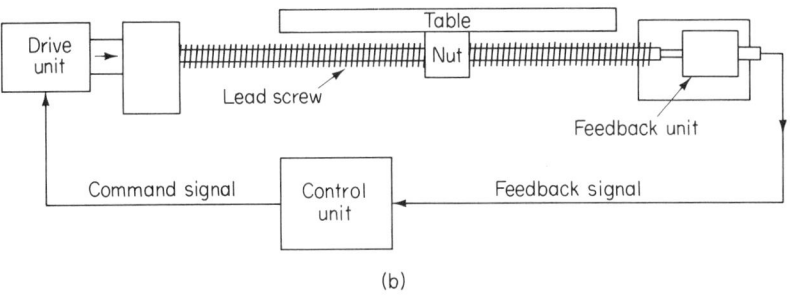

(b)

Figure 1.4

Sec. 1.3   Computer Numerical Control   7

Reeds or other forms of switching mechanisms are used to transmit signals.

1.3 COMPUTER NUMERICAL CONTROL

In computer numerical control (CNC), the memory is a computer. Essentially, a computer is the control system applied to CNC machinery. This makes it possible to store information in the computer instead of punching holes in a tape. In this manner much more information can be stored and more easily retrieved. Should conditions change, it is also easier to access the computer memory and make changes than it is to make corrections and cut a new tape.

CNC machines have manual data input (MDI) capabilities that make it possible to override existing programs. This makes it possible to change, or insert, new dimensions, feeds, speeds, and so on. CNC machines are soft wired. This means that the voltage output is wired directly to electromechanical devices such as solenoids. A computer that is soft wired to the controller is used to support a machine tool or a series of machine tools.

Paper tape (or magnetic tape, disk, or drum) is used to input information into a computer memory, which in turn operates the machine. These tapes are 1 in. wide. When punched, the holes are spaced 0.100 in. apart horizontally and vertically, as shown in Fig. 1.5. Tape materials may be paper, Mylar, or laminated paper-Mylar.

Programs may be input into computer memory directly from the keyboard of a minicomputer. A by-product may be a printout or hard copy.

Holes may be punched into paper tape for present or future use. Magnetic tapes, indicated by dashes in Fig. 1.6(c), may also be used to store programs. These are cartridges of -in.-wide magnetic tapes. Codes are magnetically stored on seven channels at speeds of about

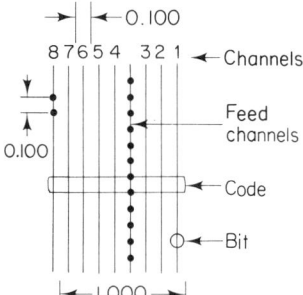

Figure 1.5

50,000 characters per second. A single reel of tape can store about 15 million characters of data.

1.4 CODES

Figure 1.6(a), (b) and (c) show three codes used in tapes: Figure 1.6(a) shows the Electronic Industries Association (EIA) code; Fig. 1.6(b), the American Standard Code for Information Interchange (ASCII) code; and Fig. 1.6(c), the code used with magnetic tape.

Modern CNC tape readers will accept either EIA or ASCII codes. Special read/write heads are needed to read magnetic codes. The EIA and ASCII codes are punched into eight vertical channels. The EIA code requires that for each character the number of horizontal holes

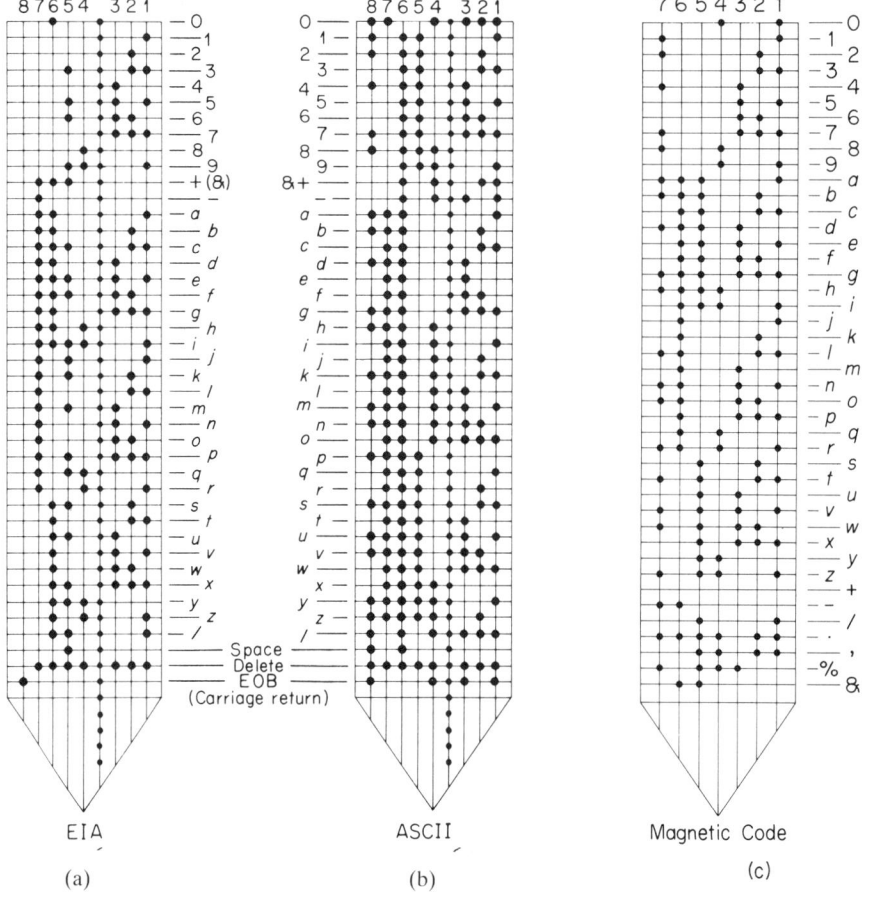

Figure 1.6

Figure 1.7 (*Courtesy of Ex-Cell-O Corporation, Remex Division*)

punched into the tape be odd. If the character has an even number of horizontal holes in a row, an extra hole must be punched into the fifth column. Thus, in Fig. 1.6(a), the character 5 has two holes and a parity hole punched into the tape. This meets the odd-parity requirement in the EIA system. For the ASCII system, Fig. 1.6(b), the character 5 has an even number of holes punched into the tape.

It should be noted that a tape reader has the capability of reading holes either mechanically or electronically. In the mechanical system a sprocket tooth enters a punched hole in the tape, causing an electrical contact, or combination of contacts, to be made. This would be the case if the parity hole is omitted, as in the EIA system. In the EIA system a photoelectric reader uses a light source on one side of the tape and a row of solid-state electronic cells lines up on the other side of the tape. Figure 1.7 shows such a photoelectric reader. It should be noted that the information is received and stored and not used until a block of information is complete. As will be seen later, the "end of block" signals the release of this information.

Another type of photoelectric light tape reader is reflective. In this system the reader is on the same side of the tape as the light source. It picks up the reflective light in the absence of a hole. the nonreflection of light indicates a hole.

The magnetic systems may be either a tape or a drum. If tape, a thin coating of magnetic material is applied to the nonmagnetic plastic surface of the tape. Floppy disks are made from nonmagnetic materials. Both sides are coated with a magnetic material. Drums are metal cylinders coated with magnetic materials.

1.5 DIRECT NUMERICAL CONTROL

Direct numerical control (DNC) eliminates the need for tapes, disks, or drums to store information. The program is typed directly from a terminal to an NC machine tool or group of machines. If the machine tool is a CNC, the signal may run the machine directly or be stored in the memory of the machine computer.

Because NC machines do not have a computer, the prevailing view some years ago was that a central computer would supply the needed signals in "real time" directly to the NC machines while the machining operations were in process, thus servicing many machine tools at once. Since many machines would be serviced by one central computer, it soon became apparent that if anything went wrong, all machines connected to the central computer would shut down. With the advent of CNC, it became possible to feed information from the large computer directly into the memory of each CNC computer. Now if the large central computer failed, the separate machine tools could rely on their own computers, and production could continue uninterrupted. The connection between the large central computer and the CNC computer was necessary only for the length of time required to process the NC information into the CNC computer.

DNC has advantages when a great deal of control information is to be processed and used to operate many machine tools, or when the programs are lengthy and complex. It is also useful in flexible manufacturing situations where many machine tools are linked in a production line. The ability of a central computer to distribute information to many machine tools has caused the acronym DNC to mean "distributed numerical control."

1.6 CAD/CAM and CIFM

Computer-aided design (CAD) makes use of the computer as a tool to produce scale layout drawings imaged on a screen. A scale model, three-dimensional, if desired, permits the operator to extract subassemblies or piece part details from the scale model. The part details are used to select tools, design fixtures, indicate clamping positions, locate blocks, and even generate bills of materials. It is also possible to use programs that will input customer requirements and to monitor design analyses such as material strength and assembly tolerances and allowances. The evaluation of tolerances and allowances before manufacturing takes place is very important to successful computer-integrated manufacturing (CIM) systems.

Once the CAD cycle is completed, the geometric model is sent to the numerical control department, where the machining operations and sequences are established. The surfaces to be machined are determined and digitized to establish the paths the cutter are to take, and the tools and fixtures are determined.

Having put all the foregoing information into the computer, it

Sec. 1.6   CAD/CAM and CIFM   11

produces the program to be used. This program is "dry run" to determine errors, after which it is released for production to CAM.

It should be noted that quality control programs may be inserted into CAD/CAM systems to inspect parts automatically as they are produced. If the quality control program detects an error, it is able to calculate the needed correction and to transmit the correction to the CAM computer, which then makes the adjustment. The design department, the manufacturing facility, and the assembly department are notified of the changes. The marriage of CAD and CAM results in the next step, computer-integrated manufacturing (CIM). The steps

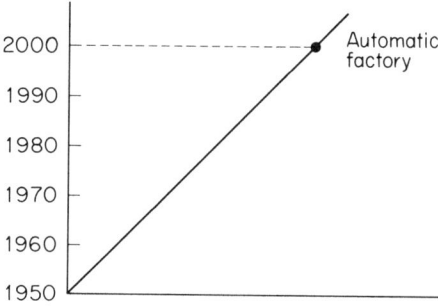

Figure 1.8(a)

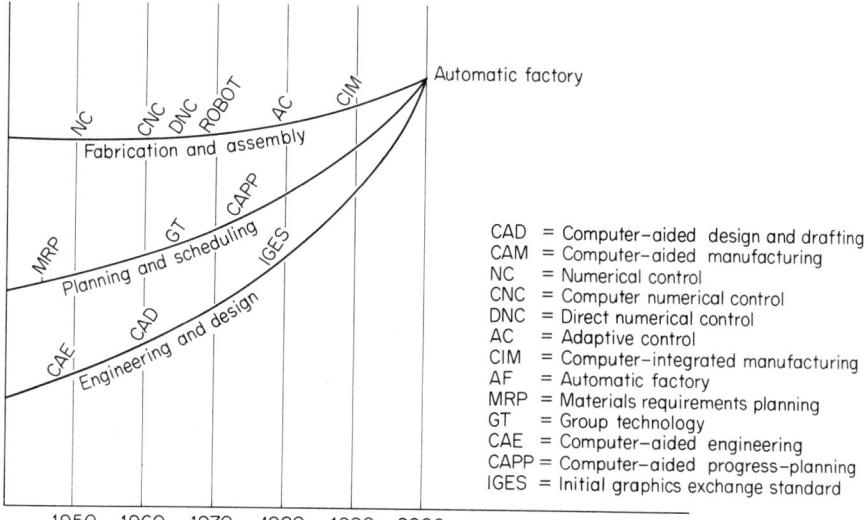

CAD = Computer-aided design and drafting
CAM = Computer-aided manufacturing
NC   = Numerical control
CNC = Computer numerical control
DNC = Direct numerical control
AC   = Adaptive control
CIM = Computer-integrated manufacturing
AF   = Automatic factory
MRP = Materials requirements planning
GT   = Group technology
CAE = Computer-aided engineering
CAPP = Computer-aided progress-planning
IGES = Initial graphics exchange standard

Figure 1.8(b)

above are linked so that the movement from operation to operation is smooth for completion of the end product.

Flexible manufacturing systems (FMS) or computer-integrated flexible manufacturing (CIFM) is the coupling of automation and control of the manufacturing processes from receipt of the raw materials through the design, completion, and shipment of the completed part or assembly. Thus the control of parts, storage, and transportation from station to station, pallets, tools, fixtures, machine tools, robots, packaging, and shipping are all integrated into one automated effort.

Once the CAD portion is completed, the parts programmers and checkers produce the program. Tools are selected as described in the program. The part to be machined is usually in a fixture that is fastened to a pallet. The pallet is mounted on a machine tool and the first operation is machined and inspected. The pallet and partially machined part are moved through the various machines for processing, or to the shipping department.

To succeed, all segments of FMS must be integrated and each step in the process must be controlled. Tools, machines, and material storage, together with quality control, must be accurately controlled. Figure 1.8(a) projects the development toward the creation of the automatic factory. Figure 1.8(b) delineates this development.

1.7 AUTOMATIC ADAPTIVE CONTROL

Adaptive control, by definition, is the process through which parameters are automatically optimized, or constrained, to achieve maximum production and dictated quality. As an example, feeds and speeds may be automatically adjusted to produce optimum tool performance, or feeds and speeds may be constrained to enhance surface finish requirements. Also, other factors may need to be either enhanced or constrained to produce dictated results: factors such as tool materials, workpiece materials, depth of cut, feeds and speeds, cutter life, and cutter geometry.

Automatic adaptive control (AAC) monitors, evaluates, and corrects performance as it happens. This process is still being refined. It will be the centerpiece of the automatic factory. In the near future, FIM and AAC will develop and make the automatic factory a reality (see Fig. 1.8).

Figure 1.9 is an idealized block diagram of a closed loop that monitors an AAC process. The censors that measure performance accept data from the input and output end of the controller and from the output end of a designated process. It measures and evaluates this information and sends the evaluation to the controller through a

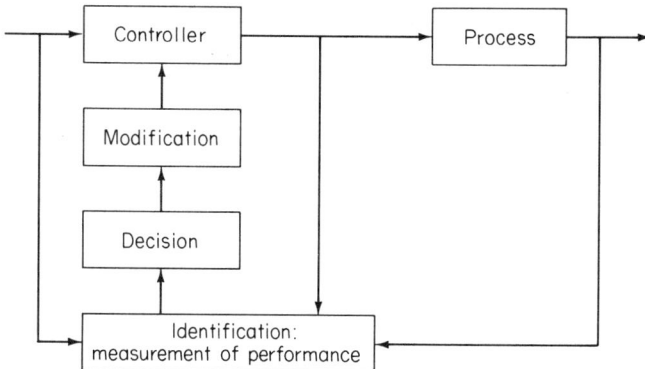

Figure 1.9

decision-making section and a modification section. Accordingly, corrections are made in the process section.

This type of control could take place at a pallet station, a machine tool, an inspection or shipping station, and so on. In an automatic factory it would take place at every phase through which the product 9passes. The goal is to control every facet of the manufacturing effort with computers. Computer control is to replace control by human beings.

QUESTIONS AND PROBLEMS

1.1. Trace the development of the control process of manufacturing from early times to the present.

1.2. What roll did John Parsons play in the development of NC?

1.3. What is ROM?

1.4. Describe NC, CNC, and DNC.

1.5. (a) Define linear interpolation.
   (b) Define circular interpolation.

1.6. (a) What is "bubble memory"?
   (b) How does it advance automation?

1.7. Trace the power train on a conventional lathe and explain the process by which it removes metal.

1.8. Repeat Prob. 1.7 for the milling machine.

1.9. Repeat Prob. 1.7 for the turret lathe.

1.10. List the effect of controls (electrical, mechanical, etc.) on mass production.

1.11. Define "open loop" as it applies to production machines. Explain each of the units in Fig. 1.4(a).

1.12. Define "closed loop" as it applies to production machines. Explain each of the units in Fig. 1.4(b).
1.13. How does a "tape" control the movement of a cutting tool or machine table?
1.14. (a) What is the control mechanism for a numerical control machine tool?
(b) What are its advantages?
1.15. (a) What is MDI?
(b) What are its advantages?
1.16. Explain the term "soft wiring" as it relates to the computer.
1.17. Explain the many ways that a computer may be programmed as it relates to machine tool control.
1.18. (a) How does a paper tape feed a signal into a computer?
(b) Electronic system may operate in one of two ways with NC. Explain them.
1.19. Describe the use of tapes, drums, and disks when used to store data.
1.20. (a) What is DNC?
(b) What are the disadvantages of connecting a series of NC machines to a computer?
(c) What are the advantages when several CNC machines are connected to a central computer?
1.21. (a) What is CAD?
(b) Describe some of its capabilities.
1.22. Repeat Prob. 1.21 for CAM.
1.23. What is CIM?
1.24. What is CIFM?
1.25. Define "adaptive control." Give at least one example other than the one given in the text.
1.26. (a) What is "automatic adaptive control"?
(b) How does it differ from "adaptive control"?
1.27. Describe the operation shown in Fig. 1.9.
1.28. Draw a block flow diagram of what you think an automatic factory would look like. Label all parts.

2

Computer Mathematics

2.1 CIRCLE AND STRAIGHT LINE

Circles may be divided in many ways. Two of the most common divisions used are degrees and radians. First let us deal with the degree method for dividing a circle.

The method of degrees divides a circle into 360 parts, each part equal to 1°. One complete circle equals 360° [see Fig. 2.1(a)]. *Straight lines* that bisect the circle (DAE) in Fig. 2.1(a) contain 180 parts. Each part is equal to 1°.

In Fig. 2.1(b), starting at D on the circle, allow line AD to rotate counterclockwise to B. This generates an angle at A. If D rotates about A to E, as in Fig. 2.1(a), the rotation has moved through 180 divisions and has generated a straight line. The angle generated is 180°. In Fig. 2.1(c), a line BC is drawn perpendicular to AD from B. This creates a triangle ABC and an angle alpha (α).

If the circle is divided into four equal parts, or quadrants, each quadrant equals 90°. The horizontal axis will arbitrarily be designated as the X axis and the vertical axis the Y axis. For the present we will deal with quadrant 1 and designate all values positive.

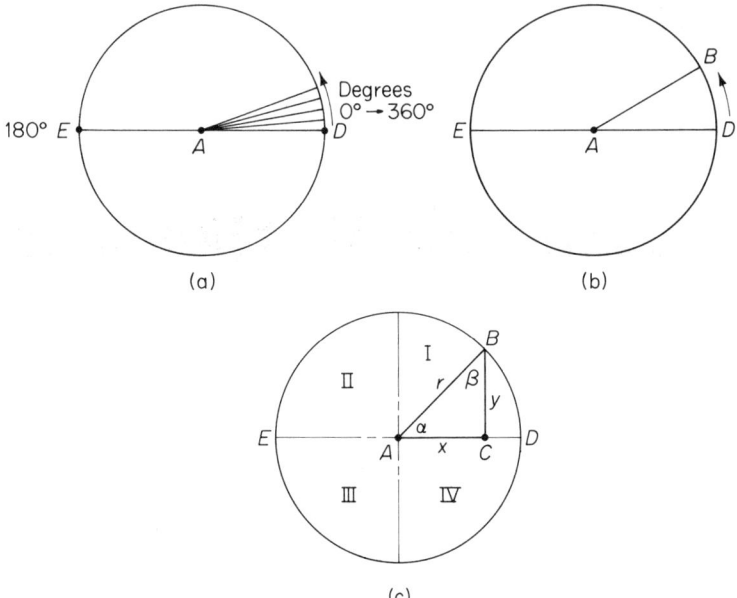

Figure 2.1

2.2 RIGHT TRIANGLE

In Fig. 2.2 the sum of the angles is

$$90° + \alpha° + \beta° = 180°$$

If the angle at $C = 90°$, the sum of the other two angles, α and β, must equal 90°.

The sides opposite the angles are called the corresponding sides. There is a fixed relationship between these angles and the length of their corresponding sides. Assuming that the length of side r is constant, then as the length of side y becomes shorter, the length of x must get longer. Also, if side y gets shorter, its corresponding angle α must get smaller.

If the sum of the angles α and β is equal to 90°, then as angle α gets smaller, angle β must get larger.

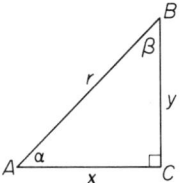

Figure 2.2

Sec. 2.2 Right Triangle

These relationships are reflected in six trigonometric relationships:

$$\text{sine } \alpha° = \sin \alpha° = \frac{\text{opposite side}}{\text{hypotenuse}} = \frac{y}{r}$$

$$\text{cosine } \alpha° = \cos \alpha° = \frac{\text{adjacent side}}{\text{hypotenuse}} = \frac{x}{r}$$

$$\text{tangent } \alpha° = \tan \alpha° = \frac{\text{opposite side}}{\text{adjacent side}} = \frac{y}{x}$$

These three functions are usually enough to solve most problems. In addition to the sine, cosine, and tangent, the reciprocal functions are available. They are:

$$\text{secant } \alpha° = \sec \alpha° = \frac{\text{hypotenuse}}{\text{adjacent side}} = \frac{r}{x}$$

$$\text{cosecant } \alpha° = \csc \alpha° = \frac{\text{hypotenuse}}{\text{opposite side}} = \frac{r}{y}$$

$$\text{cotangent } \alpha° = \cot \alpha° = \frac{\text{adjacent}}{\text{opposite side}} = \frac{x}{y}$$

Another relationship that is used when solving triangles is the Pythagorean triangle. This states that the sum of the square of the two sides of a triangle is equal to the square of the hypotenuse, or

$$r^2 = x^2 + y^2$$

This says that if two sides of a triangle are known, the third side may be found.

Example 1

Calculate the missing parts in Fig. 2.3.

Solution (a) The sum of $\alpha + \beta = 90°$. Therefore,

$$\beta° = 90° - \alpha° = 90° - 50° = 40°$$

(b) To find y, use the sine relationship, since an angle (50°) and the hypotenuse of the triangle are known.

$$\sin 50° = \frac{\text{opposite}}{\text{hypotenuse}} = \frac{y}{10}$$

Referring to the trigonometry tables, or using a calculator,

$$\sin 50° = 0.766$$

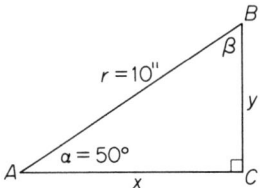

Figure 2.3

Solve for *Y*:

$$y = 10 \sin 50° = 10(0.766) = 7.66$$

(c) To find *x*, use the cosine relationship, since the angle 50° and the hypotenuse (10) are known.

$$\cos 50° = \frac{\text{adjacent}}{\text{hypotenuse}} = \frac{x}{10}$$

Therefore,

$$x = 10 \cos 50° = 10(0.643) = 6.430$$

(d) Use the Pythagorean theorem to verify the result.

$$r^2 = x^2 + y^2 = 6.430^2 + 7.660^2 = 100^2$$

$$r = \sqrt{100} = 10 \quad \text{(check)}$$

2.3 SPECIAL RELATIONSHIPS

There are several special relationships that may be used to illustrate the use of the formulas stated in Section 2.2.

45° Triangle

Example 2

Given a 45° right triangle [Fig. 2.4] to develop the sine, cosine, and tangent functions for angle alpha (α).

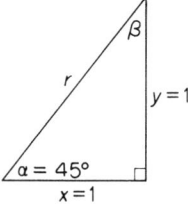

Figure 2.4

Sec. 2.3 Special Relationships

Solution Since angle $\alpha = 45°$, angle β must also equal $45°$.

$$\beta = 90° - 45° = 45°$$

Since both angles are equal (45°), the sides opposite the hypotenuse are also equal [see Fig. 2.4]. Therefore, if $x = 1.000$, then $y = 1.000$. The hypotenuse is

$$r^2 = x^2 + y^2$$
$$r = \sqrt{x^2 + y^2} = \sqrt{1.000^2 + 1.000^2}$$
$$= \sqrt{2.000} = 1.414$$

and

$$\sin 45° = \frac{y}{r} = \frac{1.000}{1.414} = 0.707$$

$$\cos 45° = \frac{x}{r} = \frac{1.000}{1.414} = 0.707$$

$$\tan 45° = \frac{x}{y} = \frac{1.000}{1.000} = 1.000$$

30°–60°–90° Triangle

Another useful relationship is the 30°–60°–90° triangle [Fig. 2.5(a)]. In this triangle the side opposite the hypotenuse is equal to one-half the length of the hypotenuse. So if r equals 2.000, y equals 1.000 [see Fig. 2.5(b)].

Example 3

(a) Solve the 30°–60°–90° triangle [Fig. 2.5(a)] for all sides.
(b) Develop the sine, cosine, and tangent angle functions for the 30° angle.
(c) Using the 60° angle, develop the sine, cosine, and tangent functions.

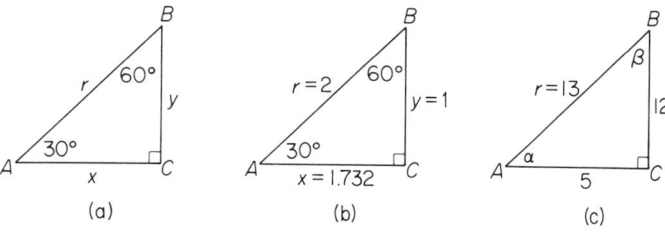

Figure 2.5

Solution (a) For the 30° angle in Fig. 2.5(b), if $r = 2.000$, 1.000. Using the Pythagorean equation gives

$$r^2 = x^2 = y^2$$

and

$$x = \sqrt{r^2 - y^2} = \sqrt{2^2 - 1^2} = \sqrt{3}$$
$$= 1.732$$

(b) The functions for the 30° angle in Fig. 2.5(b) are

$$\sin 30° = \frac{1.000}{2.000} = 0.500$$

$$\cos 30° = \frac{1.732}{2.000} = 0.866$$

$$\tan 30° = \frac{1.000}{1.732} = 0.577$$

(c) The functions for the 60° angle in Fig. 2.5(b) are

$$\sin 60° = \frac{1.732}{2.000} = 0.866$$

$$\cos 60° = \frac{1.000}{2.000} = 0.500$$

$$\tan 60° = \frac{1.732}{1.000} = 1.732$$

(d) The verification is

$$r^2 = y^2 + x^2 = 1.732^2 + 1.000^2$$
$$= 4.000$$
$$r = 2.000$$

5°–12°–13° Triangle

Another useful relationship is the 5°–12°–13° triangle.

Example 4

(a) Use the 5°–12°–13° values to verify this relationship.
(b) Develop all angles for this triangle.

Solution (a) From Fig. 2.5(c):

$$r^2 = 5^2 + 12^2 = 25 + 144 = 169$$
$$r = 13 \quad \text{(check)}$$

Sec. 2.4   Law of Sines

(b) The angles are

$$\sin \alpha = \frac{12}{13} = 0.923$$

$$\alpha = 67.4°$$

Then

$$\beta = 90° - 67.4° = 22.6°$$

2.4 LAW OF SINES

If a triangle [Fig. 2.6(a)] does *not* have an angle of 90°, the triangle must be solved using other processes. In such a triangle the sum of the angles is still 180°. The relationships between the sides and their corresponding angles are

$$\frac{a}{\sin A°} = \frac{b}{\sin B°} = \frac{c}{\sin C°}$$

These relationships are known as the *law of sines*.

Example 5

In Fig. 2.6(b) solve for:

(a) Angle $C°$.
(b) Side a.
(c) Side c.

Solution (a) Angle $C°$ is

$$C° + B° + A° = 180°$$

$$C° = 180° - B° - A° = 180° - 75° - 50°$$

$$= 55°$$

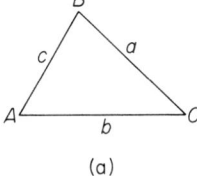
(a)

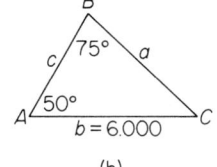
(b)

Figure 2.6

(b) Side a is

$$\frac{a}{\sin A°} = \frac{b}{\sin B°}$$

$$\frac{a}{\sin 50°} = \frac{6.000}{\sin 75°}$$

$$a = \sin 50° \frac{6.000}{\sin 75°} = \frac{6.000 \times 0.766}{0.966}$$

$$= 4.758$$

(c) From the law of sines [Fig. 2.6(b)], side c is

$$\frac{c}{\sin 55°} = \frac{6.000}{\sin 55°}$$

$$c = \frac{6.000 \sin 55°}{\sin 75°} = \frac{6.000 \times 0.819}{0.966}$$

$$= 5.087$$

2.5 LAW OF COSINES

In Fig. 2.7(a) it is not possible to solve this problem using the law of sines. The method that should be used is the *law of cosines*. This law states that

$$a = b + c - 2bc \cos A°$$
$$b = a + c - 2ac \cos B°$$
$$c = a + b - 2ab \cos C°$$

Example 6

Solve the triangle [Fig. 2.7(b)] for:

(a) Side a.
(b) Angle $B°$.
(c) Angle $C°$.

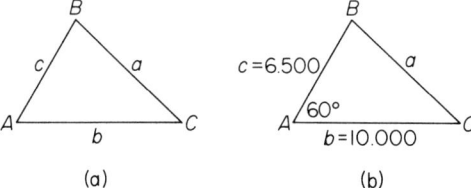

(a) (b) **Figure 2.7**

Sec. 2.6 Compensation Applications: The Milling Machine

Solution (a) This triangle *cannot* be solved by the law of sines. However, using the law of cosines to solve for side *a*, the law of sines may be used to solve for the missing parts. Using the law of cosines, solve for *a*.

$$a^2 = b^2 + c^2 - 2bc \cos A°$$
$$= 10.000^2 + 6.500^2 - 2(10.000)(6.500)\cos 60°$$
$$a = \sqrt{77.250}$$
$$= 8.789$$

(b) The law of sines can now be used to solve for either angle, $B°$ or $C°$. Solving for angle $B°$ in Fig. 2.7(b), we obtain

$$\frac{10.000}{\sin B°} = \frac{8.789}{\sin 60°}$$

$$\text{sine } B° = \frac{10.000 \text{ sine } 60°}{8.789}$$

$$B° = 80.18°$$

(c) Angle $C°$ is

$$60° + 80.18° + C° = 180°$$
$$C° = 180° - 60° - 80.18°$$
$$= 39.82°$$

2.6 COMPENSATION APPLICATIONS: THE MILLING MACHINE

A procedure known as *tool nose radius (TNR) compensation* programming is presented in Chapter 13. The *mathematics* necessary to accomplish TNR is presented below.

90° Tool Movement

In Fig. 2.8, in order for the milling cutter to finish the X and Y surfaces, it must be displaced in the *c* direction—in this case in the 45° direction. This is accomplished by displacing the cutter in the direction shown in Fig. 2.8(a).

$$x \text{ correction} = 0.500$$
$$y \text{ correction} = 0.500$$

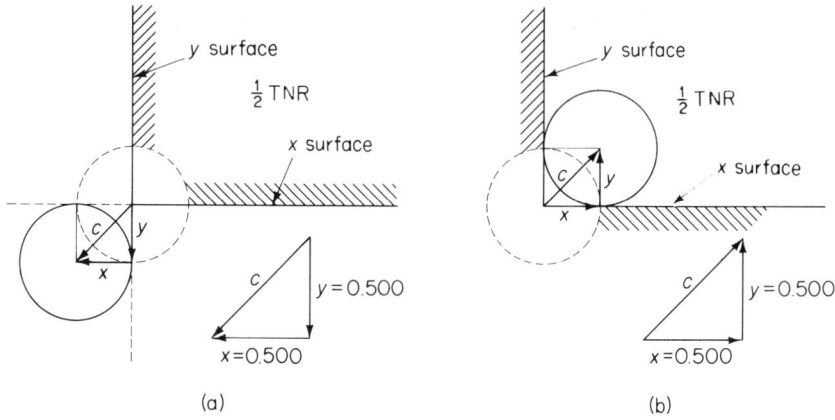

Figure 2.8

In Fig. 2.8(b), the displacement is also along the diagonal c through movements in the x and y directions. The corrections are

$$x \text{ correction} = 0.500$$
$$y \text{ correction} = 0.500$$

Tool Movements Other Than 90°

Figure 2.9 shows four basic movements that are not right-angle movements. The arrows show the direction of displacement along a hypotenuse of a triangle. This movement is accomplished along the x and y axes of the triangle generated.

In Fig. 2.10(a) the center of the cutter will move from the intersection of the two planes—the plane just machined and the plane to be machined—to the points of tangency [Fig. 2.10(b)] of the cutter and the surface of the work to be machined. This will generate the triangle shown in Fig. 2.10(c) and enlarged in Fig. 2.10(d) and (e).

If the movement is internal, as shown in Fig. 2.11, the compensation triangle ABC is generated. Figures 2.9, 2.10, and 2.11 must be thoroughly understood. There are additional compensating movements that will be studied in later chapters.

Example 7

If the angle between the work surfaces is 130° [Fig. 2.12(a)], the *alternate interior angle is also 130°*. Half that angle will be 65°, as shown in Fig. 2.12(c). The corresponding angle will be 25°, as shown in Fig. 2.12(b).

Sec. 2.6 Compensation Applications: The Milling Machine 25

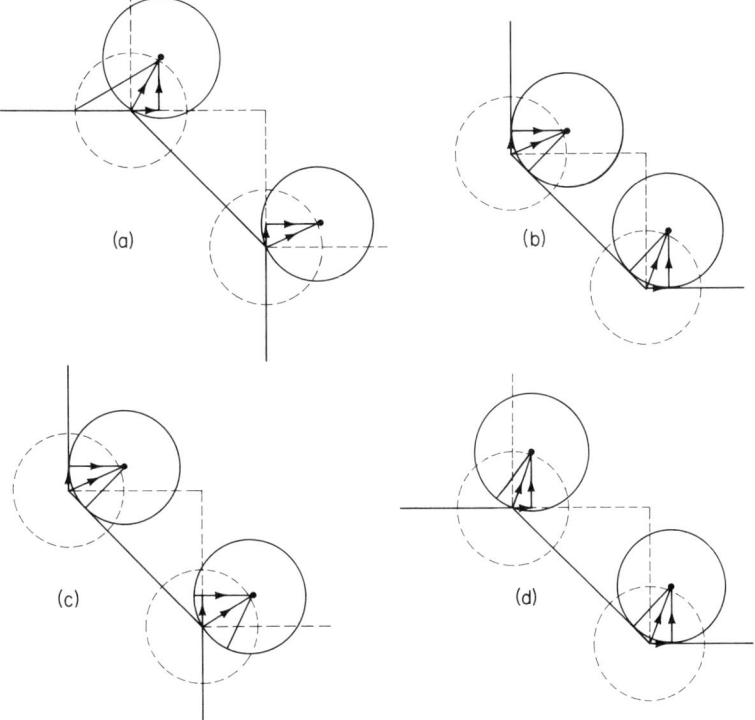

Figure 2.9

Calculate:

(a) The correction angle.

(b) The x_c and the y_c corrections.

Solution (a) Since the radius is known and the angles are known, the x correction may be calculated. Thus

$$\tan \alpha = = \frac{x_c}{r}$$

The general equation is

$$\alpha = 90 - \frac{130}{2}$$
$$= 90 - 65$$
$$= 25°$$

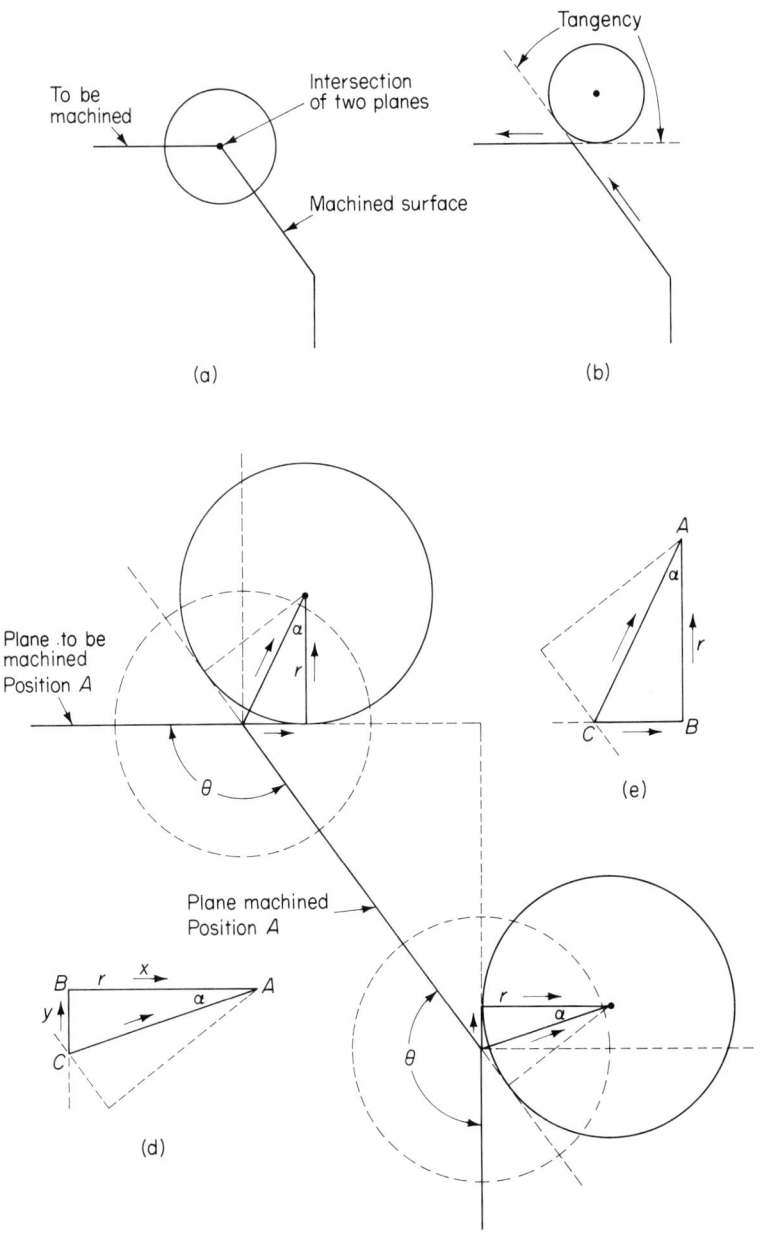

Figure 2.10

Sec. 2.6 Compensation Applications: The Milling Machine   27

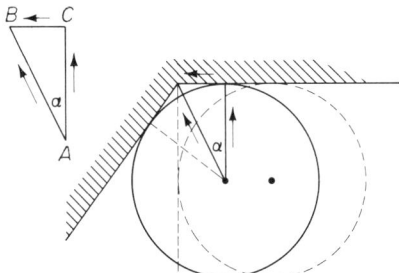

Figure 2.11

Example 8

Given a cutter TNR equal to $\frac{1}{4}$ in. Calculate the corrections in the x and y directions in Fig. 2.13.

Solution   The correction angle is

$$\alpha = 90° - \frac{180° - 75°}{2} = 37.5°$$

The y correction is

$$y_c = r = 0.250$$

The x correction is

$$\tan 37.5° = \frac{x_c}{y_c} = \frac{x_c}{0.250}$$

$$x_c = 0.250 \tan 37.5° = 0.147$$

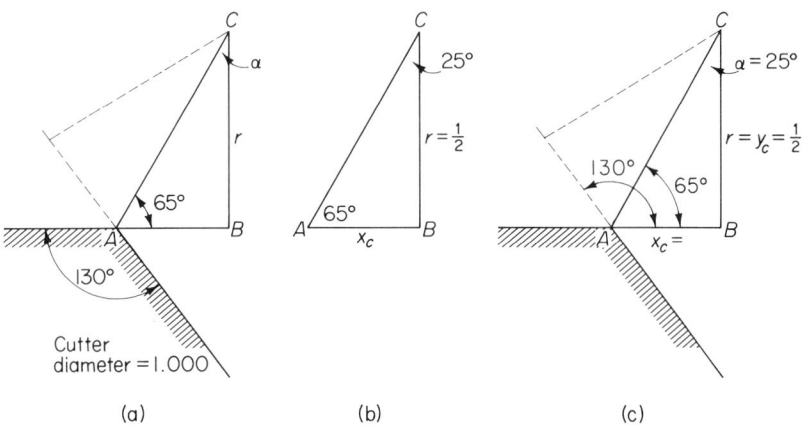

Figure 2.12

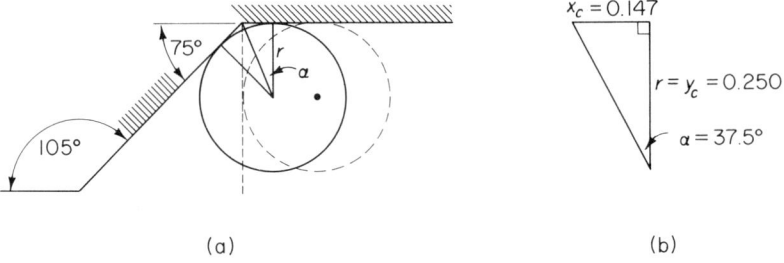

(a)   (b)

Figure 2.13

QUESTIONS AND PROBLEMS

2.1. Describe the relationship of the length of the sides and the size of the angles in a right triangle.

2.2. State the six trigonometric functions useful in the solution of right triangles.

2.3. State the Pythagorian theorem and explain the relationships of the sides to the angles in a right triangle.

2.4. Calculate the missing parts in Fig. 2.14.

2.5. Repeat Prob. 2.4 for Fig. 2.15.

Figure 2.14   Figure 2.15

2.6. Repeat Prob. 2.4 for Fig. 2.16.

2.7. Using the information supplied in Fig. 2.17, solve for:
(a) r, using the Pythagorian theorem.
(b) The angles α and β.

Figure 2.16   Figure 2.17

2.8. Given a 45° triangle, solve for the missing parts if the hypotenuse is 2 in. long.

Chap. 2   Questions and Problems   29

2.9. Given a 30°–60°–90° triangle, solve for the missing parts if the hypotenuse is 6 in. long.

2.10. Repeat Prob. 2.9 if angle A is 35° and side c is 8.000 and side b is 12.000 in. in Fig. 2.18.

2.11. Calculate the missing parts in Fig. 2.19 using the law of sines.

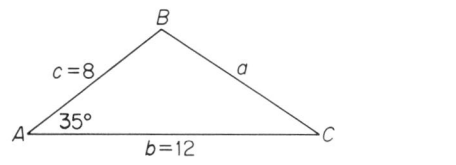

Figure 2.18   Figure 2.19

2.12. Calculate the missing parts in Fig. 2.19 using the law of cosines.

2.13. Explain the four compensation diagrams in Fig. 2.9.

2.14. Use Fig. 2.10. If the TNR is $\frac{1}{4}$ in. and the angle θ is 110°, calculate the x_c and y_c corrected values.

2.15. Repeat Prob. 2.14 when the TNR is $\frac{3}{8}$ in. and the angle θ° is 135°.

2.16. Repeat Prob. 2.14 when the TNR is $\frac{1}{32}$ in. and the angle θ° is 75°.

2.17. In Fig. 2.13(a), replace angle 105° with an angle of 125° and calculate the corrected x_c and y_c values.

2.18. Repeat Prob. 2.17 with an angle of 100°.

3

Time to Machine: Forces

3.1 CUTTING SPEEDS

Cutting speed is designated by the length of the chip produced in 1 minute. Thus if a workpiece is 1 ft in *circumference* and is revolving at one revolution per minute, that means that in 1 minute the chip that is produced will produce a chip 1 ft long. In this case the cutting speed would be 1 ft/min (see Fig. 3.1). If the same workpiece rotates at 200 rev/min, it will produce a chip 200 ft long in 1 minute. The cutting speed would be designated as 200 ft/min.

Cutting speed tables list values that are ideal and represent ideal values. They assume a rigid spindle fixture and tools, proper geometry applied to the cutting tool, minimum overhang, and so on. In general, on the lathe the spindle rotates the work. On the milling machine the spindle rotates the cutting tool.

One of the main factors that causes a tool to break down is the heat generated by the contact between the work and the tool. Since the cutting speed is greatest for the largest diameter that is being cut, given a constant rev/min rate, these larger diameters will generate the greatest amount of heat. D in the equation below is therefore taken as the largest diameter cut at a particular time.

On the lathe, the largest diameter cut is the diameter of the work. It is the work diameter D. When boring a hole, the largest diameter is the diameter of the finished cut. From the definition of the cutting speed, the length of the cut in one revolution is D. If the work revolves

Sec. 3.2 Cutting Speed and Feed Tables

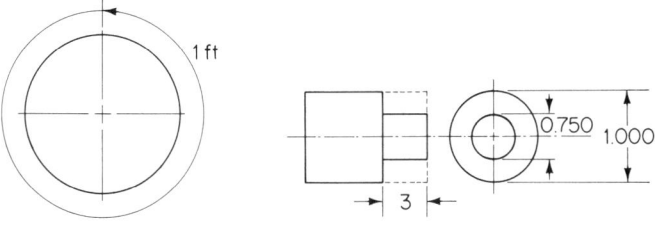

Figure 3.1   Figure 3.2

N times in *every minute*, the length of the cut will be DN. Since the diameter of the work is in units of inches and the units of cutting speed is in feet, it is necessary to divide the diameter of the work by 12. Thus

$$C_s = \frac{\pi DN}{12}$$

D = diameter of cutter, or work, in.
N = number of revolutions per minute
C_s = cutting speed, ft/min

Example 1

Given a workpiece (Fig. 3.2) to be machined. The cutter material is high-speed steel (HSS) and the workpiece material is a low-carbon steel. The cutting speed is 180 ft/min. Calculate the number of required rpm to machine the ¾-in. diameter.

Solution The cutting speed equation is

$$N = \frac{C_s \times 12}{\pi D}$$

C_s is 180 ft/min, from Table 3.1.

The rpm is

$$N = \frac{180 \times 12}{\pi \times 0.750}$$

$$= 917 \text{ rpm}$$

3.2 CUTTING SPEED AND FEED TABLES

Table 3.1 is a table of speeds and feeds that may be used as starting points when calculating the time it takes to machine a workpiece. It is important to understand that these are approximate values and that standard handbooks should be consulted for more specific values.

TABLE 3.1 FEEDS AND SPEEDS

Material	Surface Speed (ft/min)		Feeds (in./rev, 10 teeth)	
	HSS	Carbide	HSS	Carbide
Aluminum				
Wrought	700	1200	0.080	0.080
Cast	800	1200	0.080	0.080
Magnesium	800	1300	0.090	0.100
Brass	300	475	0.070	0.080
Steel				
Low Carbon	180	450	0.050	0.070
1045 ($32R_c$)	80	300	0.040	0.050
1095 ($42R_c$)	65	225	0.040	0.050
Medium and high carbon ($49R_c$)	50	100	0.010	0.010
High-carbon martensite (R_c)	--	60	--	0.010
Stainless steel				
Free machining	140	400	0.040	0.060
Cold drawn (250 BHN)	130	350	0.040	0.060
Annealed (150 BHN)	90	325	0.030	0.080
Hardened (350 BHN)	70	225	0.020	0.030
Tool steel				
Annealed (200 BHN)	100	400	0.040	0.060
Annealed (225 BHN)	55	225	0.030	0.040
Cast iron, gray	100	415	0.040	0.070
Titanium				
Com. annealed (140 BHN)	150	375	0.070	0.070
Annealed (Al-Sn)	100	275	0.050	0.070
Annealed (Al-V)	90	270	0.050	0.070
Sol. heat-treated (Al-V)	80	200	0.040	0.060
Sol. heat-treated (Al-V-Sn)	70	175	0.040	0.060
Inconel				
Nickel base, cold drawn	25	75	0.020	0.020
High-temperature alloy	12	50	0.020	0.020
Monel				
Wrought/cast	70	225	0.040	0.040

3.3 FEED RATE AND TIME TO MACHINE

Feed may be defined as the movement of the tool into the work for each revolution of the work (lathe) or the movement of the work into the tool (milling machine). The feed rate will have designated units of inches per tooth, inches per revolution, or inches per minute.

Sec. 3.3 Feed Rate and Time to Machine

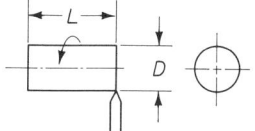

Figure 3.3

Lathe

If a tool (no radius) on a lathe moves a distance L parallel to the centerline of the work, it will move a distance fN. If the length of the work is L (Fig. 3.3), the time to machine will be

$$T_1 = \frac{L}{fN}$$

T_1 = length, in.
f = feed, in./rev
N = rev/min
L = length, in.

Drill Press

When considering a drill, the approach [Fig. 3.4(a)] must also be considered. Note that the approach distance, A_d, adds time to the machining process, so that the time equation for a drill bit becomes

$$T_d = \frac{L + A_d}{fN}$$

where T_d is the time to drill.

Example 2

Given a HSS drill with a point angle of 118°. The hole in Fig. 3.4(b) is to be drilled at 50 ft/min with a feed of 0.015 in./rev into a cast-iron block. Calculate the time required.

Solution If $C_s = \pi DN/12$, then

$$N = \frac{50(12)}{\pi(1.25)}$$

$= 153$ rev/min

C_s = 50 ft/min
L = 2.000 in.
N = 153 rev/min
f = 0.010 in./min.
A_d = 0.375 in.

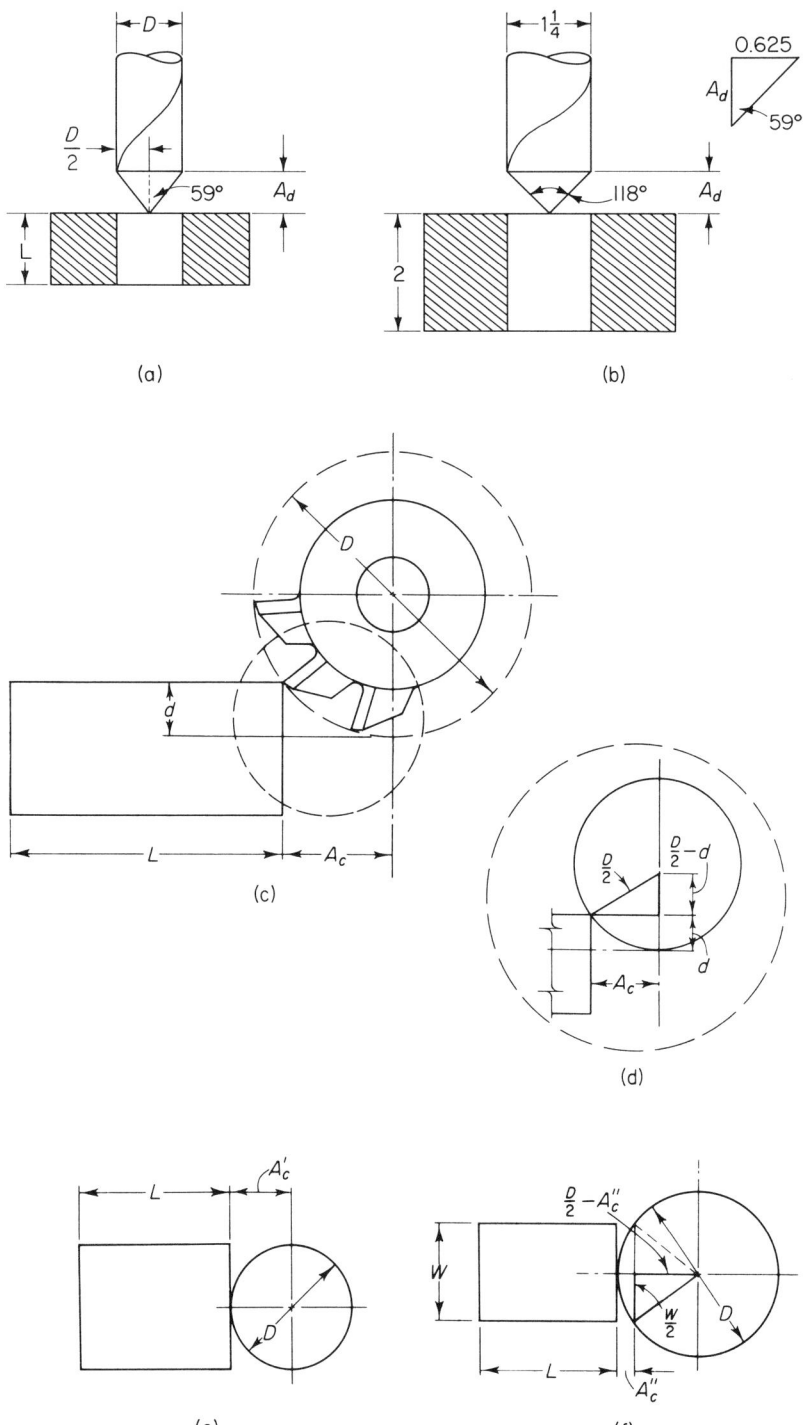

Figure 3.4

Sec. 3.3   Feed Rate and Time to Machine   35

The numerical value of the approach is

$$\tan 59° = \frac{0.625}{A_d}$$

Then

$$A_d = \frac{0.625}{\tan 59°} = 0.375$$

The time to machine the hole is

$$T = \frac{L + A_d}{fN}$$

$$= \frac{2.000 + 0.375}{0.015 \times 153}$$

$$= 1.04 \text{ min}$$

Milling Machine

Assume that a plain milling cutter is to cut a groove [Fig. 3.4(c)] into a block of material. Because of the depth of the cut, d, there will be an approach A_c. This will add to the total machining time by the amount of A_c.

From the triangle in Fig. 3.4(d),

$$\left(\frac{D}{2}\right)^2 = \left(\frac{D}{2} - d\right)^2 + (A_c)^2 \qquad \begin{aligned} A_c &= \text{approach} \\ d &= \text{depth of cut} \\ D &= \text{diameter of cutter} \end{aligned}$$

Solving for A_c gives

$$A_c = \sqrt{d(D - d)}$$

Example 3

A 4-in.-diameter steel milling cutter is to be used to cut gray cast iron as shown in Fig. 3.4(d) with a feed of 0.009 in./rev and a cutting speed of 125 in./min. If the depth of cut is to be 0.375 in. and the length of the work is 9 in., calculate:

(a) The rev/min rate.
(b) The approach.
(c) The time to take *one* cut.

Solution (a) The rev/min rate is

$$N = \frac{12C_s}{\pi D} = \frac{12(125)}{\pi(4.000)}$$

$$= 120 \text{ rev/min}$$

$D = 4.000$
$C_s = 125 \text{ ft}$
$f = 0.009$
$d = 0.375$
$L = 9.000$

(b) The approach is

$$A_c = \sqrt{d(D-d)} = \sqrt{0.375(4.000 - 0.375)}$$

$$= 1.166 \text{ in.}$$

(c) The time to take one cut is

$$T = \frac{L + A_c}{fN} = \frac{9.000 + 1.166}{0.009(120)}$$

$$= 9.413 \text{ min}$$

If a face milling cutter is used that is the same diameter, or *less than*, the width of the work, Fig. 3.4(e), the equation is

$$A'_c = \frac{D}{2}$$

Example 4

Assume the same conditions as in Example 3, except that the cutter is a face mill, as shown in Fig. 3.4(e). Calculate:

(a) The rev/min rate.
(b) The approach.
(c) The time to take one cut.

Solution (a) The rev/min rate is the same as in Example 3.
(b) The approach is

$$A'_c = \frac{D}{2} = \frac{4.000}{2} = 2.000$$

(c) The time required to take one cut is

$$T = \frac{L + A_c}{fN}$$

$$= \frac{9 + 2}{0.009 \times 120}$$

$$= 10.18 \text{ min}$$

Sec. 3.4   Horsepower Requirements

If the face mill is *large* than the width of the work, [Fig. 3.4(f)], the equation for the approach will be

$$A_c'' = \tfrac{1}{2}(D - \sqrt{D^2 - W^2})$$

Example 5

Assume the same conditions as in Example 3, except that the workpiece is 3 in. wide and the cutter is a face mill. Calculate:

(a) The rev/min rate.
(b) The approach.
(c) The time to machine one cut.

Solution (a) The rev/min rate is the same as in Example 3.

(b) The approach is

$$A_c'' = \tfrac{1}{2}(D - \sqrt{D^2 - W^2}) = \tfrac{1}{2}(4.000 - \sqrt{4^2 - 3^2}) = 0.677$$

(c)
$$T = \frac{L + A_c''}{fN} = \frac{9.000 + 0.677}{0.009(120)}$$
$$= 8.96 \text{ min}$$

3.4 HORSEPOWER REQUIREMENTS

Horsepower Requirements of a Lathe

The horsepower requirements of a lathe may be calculated as follows:

1. The cross-sectional area of a chip is the product of the depth of cut and the feed in inches per revolution. Thus

$$A = df$$

A = area of the chip, in^2
d = depth of cut, in.
f = feed, in./rev
F = force against the tool
k = constant for material
K = horsepower constant
C_s = cutting speed

2. The force of the chip on the tool is the product of the area on the chip and the constant k (Table 3.2). The constant is related to the material

TABLE 3.2 K CONSTANTS (LATHE)[a]

Material	K
Aluminum	4
Brass (medium)	6
Bronze (medium)	6
Iron	
Cast	4
Wrought	6
Malleable	4
Steel	
Low-carbon	6
Mild	8
High-carbon and alloy	10
Stainless	8

[a] Average values. Tables in handbooks should be consulted.

being cut. Thus

$$F = kA$$

3. The horsepower at the cutting edge is related to the force on the tool and the cutting speed. Thus

$$\text{hp} = \frac{FC_s}{33,000} = \frac{KAC_s}{33,000}$$

$$= KAC_s$$

where

$$K = \frac{k}{33,000}$$

4. Thus the horsepower at the cutting edge of the tool bit for a lathe may be written

$$\text{hp} = KdfC_s$$

K is a conversion constant that may be found in Table 3.2. This equation may be used to get a close approximation for single-point tools.

Example 6

Find the horsepower at the cutting edge if the material being cut on the lathe is mild steel. The depth of cut is $\frac{1}{16}$ in. (0.0625 in.). The feed is 0.012 in./rev. The cutting speed is 70 ft/min. The material constant K is 8 (From Table 3.2).

Solution

$$\text{hp} = KdfC_s = 8(0.0625)(0.012)(70)$$
$$= 0.42 \text{ hp}$$

$d = 0.0625$ in.
$f = 0.012$ in/rev.
$C_s = 70$ ft/min.

Sec. 3.4 Horsepower Requirements

The efficiency of machines ranges from 60% for 3-hp motors to about 80% for large motors. The losses result from friction that develops when the machine parts are moving. The horsepower requirements are based on the volume of metal removed per minute.

Horsepower of a Milling Machine

The horsepower requirements for *one milling cutter tooth* is

$$\text{hp}_t = \frac{V}{K'} = \frac{dWF}{K'}$$

W = width of cut, in.
d = depth of cut, in.
F = feed, in./min
f' = feed, in./tooth
N = rev/min
K' = milling constant
n = number of teeth
hp_t = hp of one tooth

The horsepower at the cutter is

$$\text{hp}_c = \frac{ndWf'N}{K'}$$

hp_c = hp of the cutter

The horsepower at the machine is

$$\text{hp}_m = \frac{\text{hp}_c}{\text{efficiency}}$$

The values of K' for a milling cutter are shown in Table 3.3.

Example 7

A 6-in. milling cutter that has 12 teeth is to machine a soft-steel surface by removing a chip 1 in. wide and $\frac{1}{4}$ in. deep with a feed of 0.005 in./tooth. Calculate the horsepower requirements of the operation if the machine is 60% efficient.

Solution The rev/min value of the cutter is

$$N = \frac{12C_s}{\pi D} = \frac{12(90)}{\pi(6)}$$
$$= 57.3$$

D = 6 in.
n = 12 teeth
d = $\frac{1}{4}$ in
W = 1 in.

TABLE 3.3 K' CONSTANTS (MILLING)[a]

Material	K'
Aluminum	2.25
Brass	1.75
Cast iron	
Soft	1.25
Medium	1.00
Hard	0.75
Malleable	1.00
Carbon steel	
Soft	0.80
Medium	0.60
Hard	0.50
Stainless	0.60

[a]Average values. Tables in handbooks should be consulted.

The horsepower requirement is

$$hp_c = \frac{ndWf'N}{K'}$$

$$= \frac{12(\frac{1}{4})(1)(0.005)(57.30)}{0.80}$$

$$= 1.1 \text{ hp}$$

$f' = 0.005$ in/T (See Table 3.1)
$K' = 0.80$
$C_s = 90$ ft/min
$e = 60\%$
$hp_c =$ hp cutter
$hp_m =$ hp machine

The horsepower requirements of the motor are

$$hp_m = \frac{1.1}{0.60} = 1.8$$

Horsepower of a Drill Press

The horsepower of a drill press is given by the equation

$$hp = K''C_sD^{0.8}f^{0.7}$$

$K'' =$ drill constant
$f =$ feed, in./rev
$D =$ diameter, in.
$C_s =$ cutting speed, ft/min

The horsepower for a drill bit is based on the ratio of the drill

TABLE 3.4 K" CONSTANTS (DRILL PRESS)

Material	K"
Cast Iron	
Soft	1.00
Medium	0.54
Hard	0.58
Steel (carbon)	
Low	1.20
Medium	1.40
High	1.70

diameter to the length of the chisel edge. The values of K'' are given in Table 3.4.

Example 8

A 1-in.-diameter drill is used with a cutting speed of 50 ft/min and a feed of 0.012 in./rev. The material is cast iron. What is the horsepower required at the drill point?

Solution

$$\text{hp} = K'' C_s D^{0.8} f^{0.7} \qquad K'' = 0.54$$
$$= 0.54(50)(1)^{0.8}(0.012)^{0.7} \qquad f = 0.012 \text{ in.}$$
$$= 1.22 \text{ hp} \qquad D = 1 \text{ in.}$$
$$\qquad C_s = 50 \text{ ft/min}$$

QUESTIONS AND PROBLEMS

3.1. (a) Explain the term "cutting speed."
   (b) Relate this to the rev/min rate.
3.2. (a) What is the effect of heat on the life of a cutting tool?
   (b) Describe the "failure process."
3.3. Define "feed."
3.4. The diameter of a soft steel shaft is 4 in. The length of the cut is 7 in. If the feed is 0.020 in./rev, calculate:
   (a) The rev/min rate.
   (b) The time to machine the length with a high-speed-steel tool bit.
3.5. An aluminum shaft $2\frac{1}{2}$ in. in diameter is to be machined with a high-speed tool bit using a feed of 0.030 in./rev. The length of cut is 10 in. Calculate the time to take one cut on the lathe.

3.6. Assume that the tool bit in Prob. 3.4 is capable of taking $\frac{3}{8}$ in. off the diameter of a shaft. If the shaft diameter is $1\frac{3}{4}$ in., how long will it take to machine the shaft, allowing 5% of the total time for positioning?

3.7. A cast-iron shaft is machined on a lathe in 1 min with one cut. The shaft is 4 in. long and 3 in. in diameter. If the feed used is 0.012 in./rev, what cutting speed was used?

3.8. Figure 3.5 shows a 3-in.-diameter soft-steel shaft. The length of cut is 8 in. If the feed is 0.012 in./rev, calculate:
(a) The rev/min rate.
(b) The time to machine the length with a high-speed-steel tool bit.

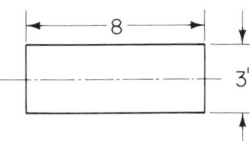

Figure 3.5

3.9. Assume a depth of cut of $\frac{1}{8}$ in. in Prob. 3.8. If the finishing depth of cut is 0.020 in., the feed is changed to 0.012 in./rev, to a feed of 0.020 in./rev, calculate:
(a) The new RPM.
(b) The time to machine.

3.10. A $1\frac{1}{4}$-in.-diameter high-speed-steel drill is used to drill 16 holes in a medium-high-speed-steel plate. The drill plate is $1\frac{1}{2}$ in. thick and the feed is 0.008 in./rev. What is the machining time to drill the 16 holes?

3.11. A $\frac{3}{4}$-in. hole is to be drilled into a cast-iron block with a feed of 0.015 in./rev. The thickness of the block is $2\frac{3}{8}$ in. and the drill bit is made from high-speed steel. Calculate:
(a) The rev/min rate.
(b) The approach.
(c) The time to machine.

3.12. A 6-in.-diameter carbide milling cutter is to be used to cut medium-hard steel, as shown in Fig. 3.4(c), with a feed of 0.008 in./rev. If the depth of cut is to be 0.050 in. and the length of cut is 8 in., calculate:
(a) The rev/min rate.
(b) The approach.
(c) The time to take one cut.

3.13. Assume the same conditions as in Prob. 3.12, except that the cutter is a face mill, as shown in Fig. 3.4(e). Calculate:
(a) The rev/min rate.
(b) The approach.
(c) The time to take one cut.

Chap. 3 Questions and Problems   43

3.14. Assume the same conditions as in Prob. 3.12, except that the work is 5 in. wide and the cutter is a face mill [see Fig. 3.4(f)]. Calculate:
  (a) The rev/min rate.
  (b) The approach.
  (c) The time to take one cut.

3.15. A 9-in. high-speed milling cutter is used to cut a groove into a piece of brass with one cut. The groove is $\frac{3}{4}$ in. deep and 10 in. long. If a feed of 0.018 in./rev is used, how long will it take to machine the groove?

3.16. The cutter in Prob. 3.15 has 24 teeth.
  (a) How many chips per minute are cut?
  (b) What is the thickness of each clip?

3.17. An 8-in. carbide face milling cutter is used to machine a soft-steel block. The cutter is the same width as the block to be machined. The feed used is 0.030 in./rev and the length of the block is 14 in. Calculate the time to take one cut.

3.18. Find the time to take one cut in Prob. 3.17 if the work is 5 in. wide.

3.19. **(a)** Find the horsepower at the cutting edge on a lathe using the conditions in Problem 3.7. The depth of cut is $\frac{1}{8}$ in.
  (b) What is the cross-sectional area of the chip?

3.20. A $1\frac{1}{2}$-in.-wide chip is removed by a 8-in.-diameter high-speed-steel milling cutter that has 18 teeth. The depth of cut is $\frac{3}{8}$ in. and the feed is 0.002 in./tooth. The material to be machined is soft steel. The operation is 52% efficient. What are the horsepower requirements of the machine?

3.21. A $\frac{3}{4}$-in.-diameter high-speed-steel drill is used to machine a hole into a medium-hardness-steel block with a feed of 0.015 in. per revolution. What horsepower is required to the drill?

3.22. Given a 4" H.S.S. milling cutter that has 16 teeth. It is to be used to cut aluminum and remove a chip $\frac{3}{4}$ in. wide by $\frac{3}{8}$ in. deep using a feed of 0.004 in./tooth. Calculate the horsepower requirements of the operation if the machine is 70% efficient.

3.23. Repeat Prob. 3.22 if a 6-in.-diameter milling cutter is used. The cutter has 24 teeth.

3.24. Given a cutting speed of 80 ft/min, a feed of 0.020 in./rev, and a drill diameter of $\frac{3}{4}$ in. The material to be cut is medium steel. Calculate the horsepower requirements if the efficiency is 65%.

4

Cutting Tools: Their Uses

4.1 THEORY OF CUTTING

The operation of a cutting tool, whether it is on a lathe, milling machine, or any other machine tool, is based on theory that is the same for all machining processes. The purpose of any operation is to achieve good surface finish at high speeds and feeds with the least effort and at the lowest cost.

The desirability of getting the maximum use from a tool before it needs regrinding is one of the objectives of tool technology. Assuming that a tool has been properly designed and made, failure may result from the wearing away of the tool's cutting edge. This changes the geometry of the tool. This geometry may be in the nature of a dull edge, roughness, or a shift in the clearance angles. Any of these changes will generate heat, which may cause the tool to lose hardness. This does not mean that the tool is soft. It simply means that the tool has softened to the point where the movement of the tool in relation to the work makes the tool soft for efficient cutting. If the relative motion of the tool to the work is too great, the tool material will rub off the cutting edge and the conditions that prevail after the rubbing away has taken place create even higher temperatures, so that larger sections of the tool's cutting edge are rubbed away. The process of softening and rubbing away continues until the tool breaks down completely.

Thus proper lubrication or cooling, sharp tools, proper angles,

Sec. 4.1 Theory of Cutting

careful selection of tool materials, proper feeds and speeds, and the proper setting up of the tool relative to the work surface all help in cutting down the heat generated and prolonging the tool life. Tool life is defined as the length of time a tool will operate before failure occurs.

Another cause for tool failure results from high stresses set up by the tool within the workpiece and within the resulting chip. The metal is said to work-harden and as a result, greater forces are needed to separate the chip from the parent metal. Some materials (stainless steel 18-8) work-harden severely, which becomes an important factor in the cutting process.

Work hardening can sometimes be circumvented by taking cuts deep enough so that the previously hardened surface of the work is removed farther up the cutting edge where the tool is stronger and creates a "chipping" effect. The same technique may be used for removing scale from hot-rolled material. Still another method is to remove the work-hardened surface by preheating the work.

Chip formation is a function of the tool bit and the nature of the material being cut. It may be classified as continuous, discontinuous, or built-up edge and is always the result of shear. The planer tool will be used to illustrate chip formation because the tool is set perpendicular to the work and operates along the length of the work.

A *continuous chip* [Fig. 4.1(a)] is obtained when cutting ductile material. The chip is severely deformed and comes off in the form of a long string or coils into a tight roll. Prior to being cut, metal is much longer than the chip being removed. Since the volume of the material before cutting and after cutting is the same, and the width of the cut remains the same, the thickness of the chip after cutting must be greater than before cutting [see Fig. 4.1(a)]. The ratio r is the ratio of the values before cutting to after cutting should be about, or less than, 1:2 to produce good results.

A *discontinuous chip*, such as results when cutting cast iron, is shown in Fig. 4.1(b). Assume that the cutting action is just starting. The material starts to slide up the face of the tool. Since the shearing forces are high, a crack develops early as the cutting proceeds. This cut fractures small pieces of material off the work. The results are discontinuous. In gray cast iron the structure of the material—long graphite stringers—means that the forces need only break through the material that separates one graphite stringer from the next.

Built-up edges may result during continuous cutting [Fig. 4.1(c)]. In this action, the high heat generated welds a small chip to the tool. As the weld builds up, the welded chip grows until it breaks away from the tool. Figure 4.1(c) shows part of the weld remaining with the work and the rest of the weld moving up the face of the tool. Neither

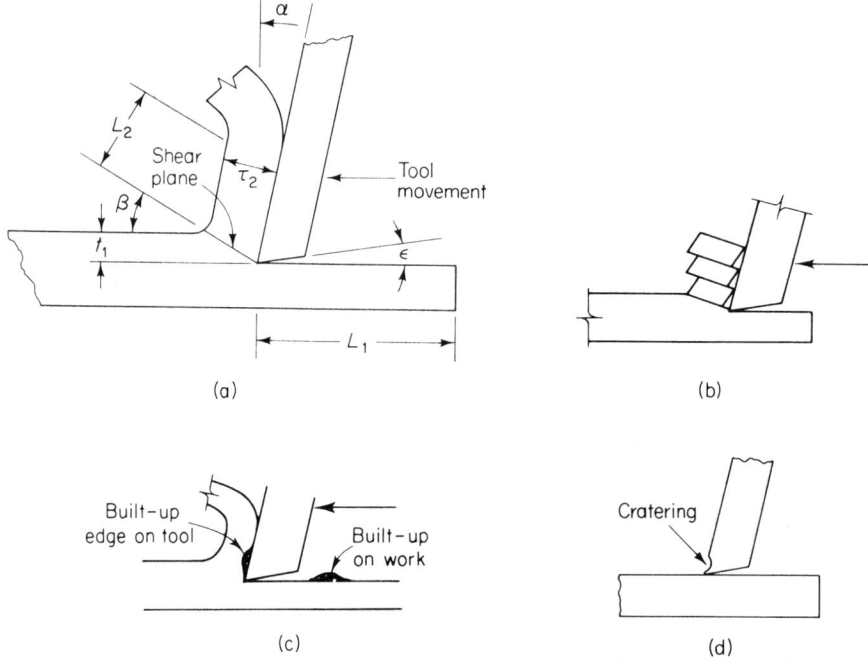

Figure 4.1

of these conditions is desirable because one leaves a rough working surface, and the other, while welded to the tool, interferes with its cutting action.

Cratering [Fig. 4.1(d)] in the face of the tool may also result from the welding mechanism. This occurs in the face of the tool. Each time the chip breaks away from the face of the tool, it takes a very small amount of material off the face of the tool. The accumulated effect of many such actions is a crater in the face of the tool.

Flank wear is a continuous process that takes place below the cutting edge, in the face of the flank of the tool. As the wear flat widens, it destroys the clearance below the cutting edge, which in turn causes rubbing, increases heat generation, increases the cutting forces against the tool, and causes greater flank wear.

The cutting section of the tool bit may be divided into three zones: Z_f, the friction zone; Z_d, the deformation zone; and Z_w, the work-surface zone. These zones are shown in Fig. 4.2(a).

The forces shown in Fig. 4.2(b) are f_{rF}, the friction force along the face of the tool and the normal force, f_{rN}. The resultant of these two forces F_r, is the force exerted by the tool on the work. The work exerts an equal but opposite force R on the tool to f_R. This resultant

Sec. 4.1 Theory of Cutting

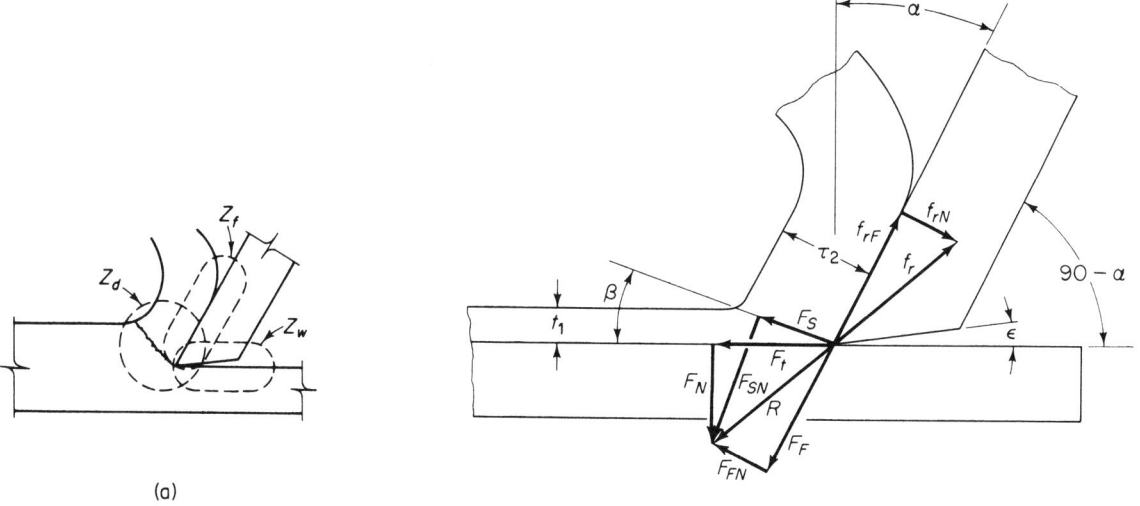

Figure 4.2

force R has two components: a normal force, F_N, to the shearing action and the shearing force, F_S. It should be pointed out that if the angle $(90° - α°)$, gets larger, the angle $β°$ [Fig. 4.2(b)] gets smaller and the forces required to cut the material gets greater.

The system of forces described above are *orthogonal* and are two-plane-dimensional. Most cutting is more complex and is three-dimensional *oblique* cutting. Orthogonal cutting is shown in Fig. 4.2(c). This type of cutting requires the tool cutting edge to be perpendicular to the direction of travel. In contrast, Fig. 4.2(d) shows an example of oblique cutting. In Fig. 4.2(c) the chip comes off the work parallel to the face of the work. In Fig. 4.2(d) the chip comes off at an angle to the work. The force generated may be measured with a strain gage. There are additional forces operating at the cutting edge that are very small and have not been considered here.

From the discussion above, it can be seen that the friction must be kept as low as possible to reduce the heat generated. This can be done with the use of lubricants that form an oily film on the surface of the metal, thus make the shearing of metal easier. This is the primary purpose of a lubricant. It may be a fatty oil, sulfurized mineral, or fatty oil.

Its second effect is to remove heat generated during the cutting operation both from the energy converted to heat during the shearing process and from the friction generated by the chip against the tool face.

When the cutting operations are continuous and the lubricant cannot remove the heat rapidly enough, water-soluble oils may be used. When mixed with a high concentration of water, the cooling effect is greatly increased with some lubricating properties retained. These mixtures do not corrode the steel parts with which they come into contact.

Thus lubricating oils are used chiefly to reduce friction and water-soluble oils are used chiefly as coolants. In general, recommended coolants and lubricants are: for steels and wrought iron, water-soluble oils or sulfur-based or soluble oils; for brass, copper, bronze, monel metals, and malleable iron, soluble oils. Cast iron is machined dry.

4.2 TOOL BIT MATERIAL

The materials used for tool bits must possess certain qualities to be effective. They must possess hardness, strength, and toughness and they must be heat resistant. There are conditions associated with the materials being cut that affect the selection of the tool material: the character and the condition of the workpiece material, the amount of

Sec. 4.2 Tool Bit Material

stock to be removed, the required provision of the finished workpiece, the surface finish, the rigidity of the workpiece and the tool holder, the condition of the machine tool, and so on. All of these items must be considered when selecting a cutting tool material. To operate efficiently, cutting tools must have properties such as hot hardness, toughness, resistance to cratering, wear resistance, and resistance to shock.

It should be noted that some hardened material retain their hardness after being subjected to high temperatures during the cutting operation and cooled to room temperature. Such tool materials are: ceramics, carbides, cast materials, and high-speed steel (HSS) up to 1000°F. Carbon steel loses its hardness almost immediately upon being cooled from elevated temperature to room temperatures.

High-Carbon Steel

Tools made from high-carbon steels, 0.80 to 1.30%, are used for small-quantity production of wood parts, or machining soft materials such as free-cutting steel or brass. It is important that the operational temperatures be kept below 400°F because the tool material starts to lose its hardness above this temperature. For this reason coolants should be used freely. Sometimes vanadium and chromium are added to give better hardenability.

High-Speed Steel

These tool materials are of several varieties. The most common type are the 18-4-1 high-speed steels. This tool material has approximately 18% tungsten, 4% chromium, and 1% vanadium as its alloying elements. Another type has about the same amount of tungsten, but about 5 to 10% cobalt added as one of the principal alloying elements. This improves the red hardness of the material, although operating at higher temperatures than the 18-4-1 steel is more expensive.

There are other types of HSS tool materials. High-speed steel will operate up to about 1100°F and retain its effectiveness. This ability to operate at high temperatures makes it possible to increase the cutting speed, or depth of cut. However, a coolant should be used freely to increase the maximum life of the tool.

Nonferrous Cast Tool Steels

This material cannot effectively, or economically, be shaped by cold or hot working and must be cast into desired shapes. It operates best at elevated temperatures (approximately 1500°F), losing efficiency when operated at low temperatures. Therefore, the cutting speed may be increased without damage to the tool. Deeper cuts may be taken at higher cutting speeds and reduced feeds. They operate at conditions somewhere between those of high-speed steels and cemented carbides.

Carbides

These are sintered materials. There are three elements usually "cemented" with cobalt as a binder: tungsten, titanium, and tantalum. Titanium and tantalum are added to the tungsten to achieve various desirable properties that tungsten alone might not have.

Tungsten carbides are used primarily to machine cast iron, nonferrous metals, plastics, rubber, and so on. Tungsten–titanium and tantalum–titanium carbides may be used to machine steels, tough materials, and most materials that may be machined with tungsten carbides. In general, the deeper the cut and the tougher the material being cut, the lower should be the Rockwell hardness of the tool material used. Higher Rockwell hardness materials may be used to take light cuts. Carbides, in general, may be used at cutting speeds that are about three times that used with high-speed-steel tools.

One of the problems encountered with carbides, and most other materials, is cratering. If carbide is coated with TiC, aluminum oxide, or titanium nitride, the cratering effect is reduced from 50% to 90% and tool wear is increased while cutting speed may be increased substantially.

Ceramic Tool Material

Ceramic tool materials, that is, aluminum oxide or silicon carbide, may be mixed with a glass binder. Sometimes the material is compacted without a binder. Under 2000°F, sintering takes place and the material becomes hard and brittle. It will withstand operating temperatures of about 2000°F without losing hardness or strength. Because of its high compressive strength and its low coefficient of resistance, the cutting speed may be about two to three times greater than those used with carbides.

Industrial Diamonds

These materials have little uses in present-day machining of metals. They may be used to machine aluminum, plastic, and hard rubber, and if used with very fine feeds and high-spindle speeds, for fine finishing of bored holes in steel. They are expensive and difficult to shape into desired forms. However, they are used as compacted diamond materials to cut plastics, nonferrous materials, and other soft-metallic materials. Borazon is a material with a hardness value approaching that of diamond. It is less shock resistant than diamond and withstands high temperatures.

4.3 TERMINOLOGY, CLEARANCE, RAKES, AND CHIP BREAKERS

The principles that apply to single-point tools also apply to other types of cutting tools. Probably the most important single aspect of any type of cutting tool is the fact that the cutting edge must be free to cut from either the tool itself or from the work. The question of how much support is needed, for the cutting edge depends on the type of material in the cutting tool, type of material being cut, feed, cutting speed, rigidity, and so on. The full implication of this paragraph will become more evident as this presentation unfolds.

There are several types of single-point tools available. They are solid tools that fit directly into a holder [Fig. 4.3(a)], solid tools that fit into a tool post [Fig. 4.3(b)], tool inserts that are clamped into a solid piece of material [Fig. 4.3(c)], or tool inserts that are brazed to a tool shank. Figure 4.4 shows several configurations of inserts that are available commercially.

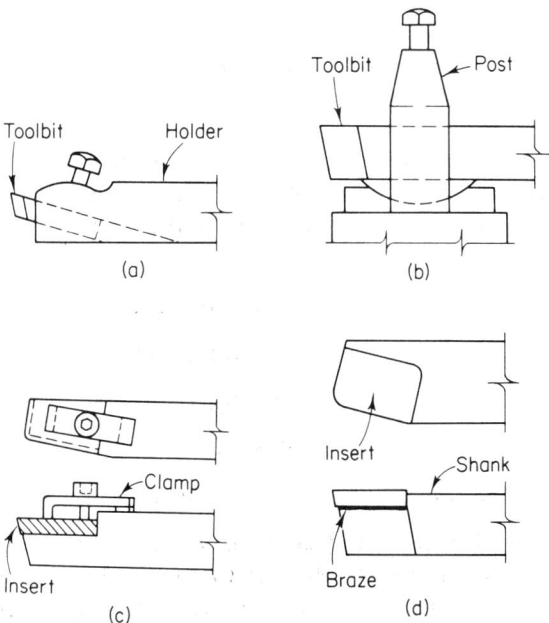

Figure 4.3

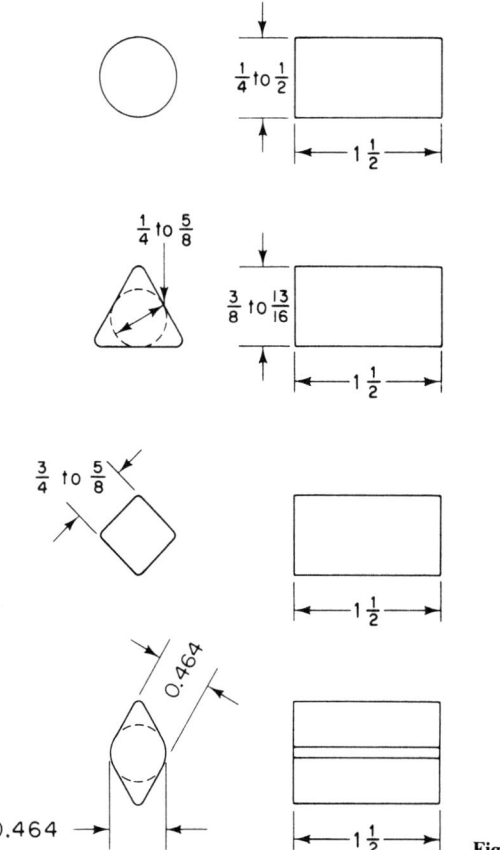

Figure 4.4

Figure 4.5(a) shows the tool bit in Fig. 4.5(b) in operation. Figure 4.5(b) shows three views of a tool bit. The following notations indicate the terminology of the various parts of the tool bit in Fig. 4.5(b).

a = back rake angle f = side cutting-edge angle
b = side rake angle g = cutting angle
c = front relief angle h = lip angle
d = side relief angle i = nose angle
e = end cutting-edge angle j = nose radius

In Fig. 4.5(b) the tool is ground with a *front relief angle c* and an *end cutting-edge angle e* so that only the tip of the tool touches the work. Since the feed is from left to right in the front view [section

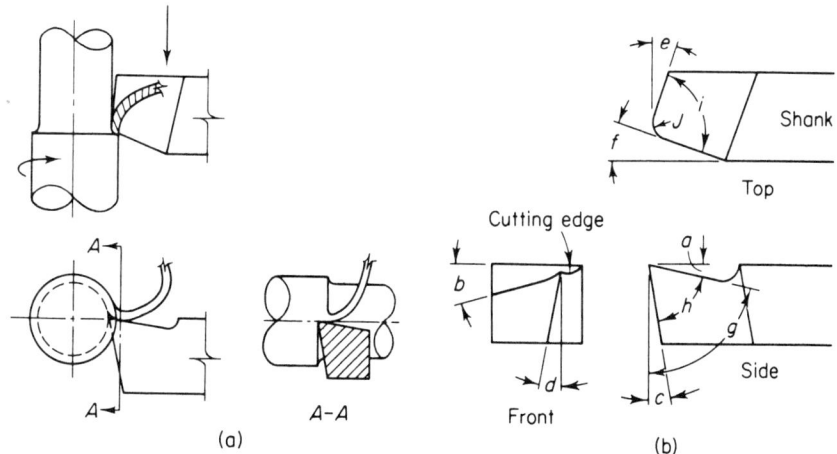

Figure 4.5

A-A, Fig. 4.5(a)], the tool must be ground with a *side relief angle d* so that the tool material below the cutting edge does not interfere with the cutting action.

The *back rake angle a*, the *side rake angle b*, and the *side cutting-edge angle f* are ground to produce the desired effect during cutting. All remaining angles (h, g, i) result from the manner in which the former angles are applied to the tool bit.

In general, the clearance and rake angles in Table 4.1 and shown in Fig. 4.6 are recommended. The relief angles for the following are: a lip angle of

$$90° - 10° - 6° = 74°$$

a cutting angle of

$$90° - 10° + 0° = 80°$$

a nose angle of

$$90° - 12° + 0° = 78°$$

Table 4.1 shows some recommended relief and rake angles for high-speed-steel tools. Angles and rakes for carbide tools should be controlled carefully. These are shown in the solid carbide tool bit (Fig. 4.7). Inserts are also used. They have been standardized and are shown in Fig. 4.8

Side relief angles should be ground consistent with the feed used and the material to be cut. In general, the side relief and the front relief should be kept as small as possible. It should be remembered that carbide is brittle and the more support the tip gets from the body

TABLE 4.1 Clearance and Rake Angles

Front relief angle	6°
Side relief angle	7°
Back rake angle	10°
Side rake angle	15°
Nose radius	$\frac{1}{16}$ in.
End cutting-edge angle	12°
Side cutting-edge angle	0°

of the tool, the less likely it is that the tip of the tool will break or chip. Front and side relief angles are ground to between 3 and 10°, with most tools ground to 5°. The softer the tool material, the greater the angle permitted. The shank side relief angle should be about 2° greater than the tip side relief angle.

A side cutting-edge angle and edge-cutting angle of about 15° should be satisfactory for most turning operations. This produces a shearing action as the tool proceeds along the work and provides a 70° nose angle. Another effect of the side cutting-edge angle is that the solid part of the cutting edge contacts the work first. This heavier cross of the tool bit is better able to take the first shock of the cutting action than the tip of the tool.

Rakes are used to aid the shearing action of the tool as the work revolves onto the tool surface. However, it is important to recognize

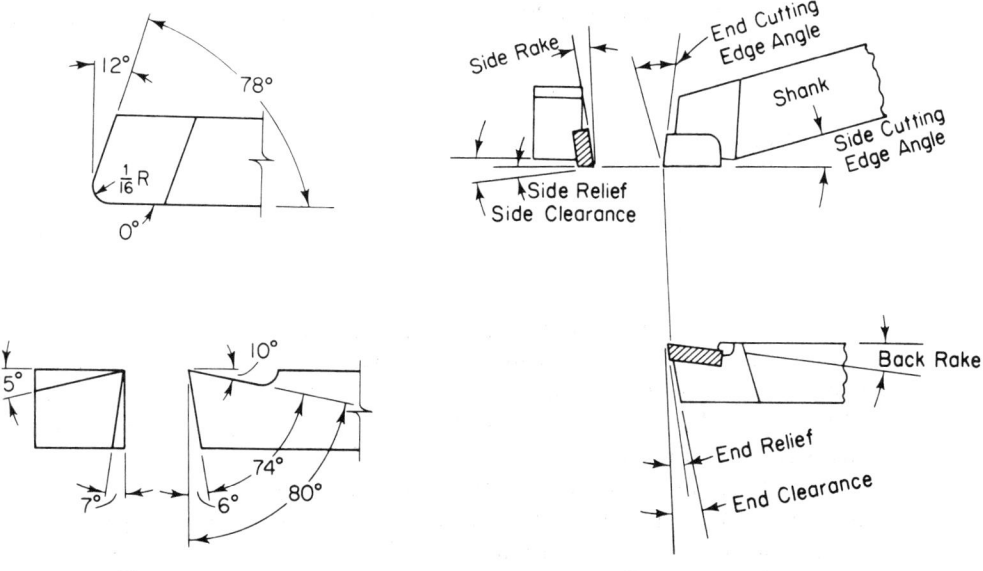

Figure 4.6 **Figure 4.7**

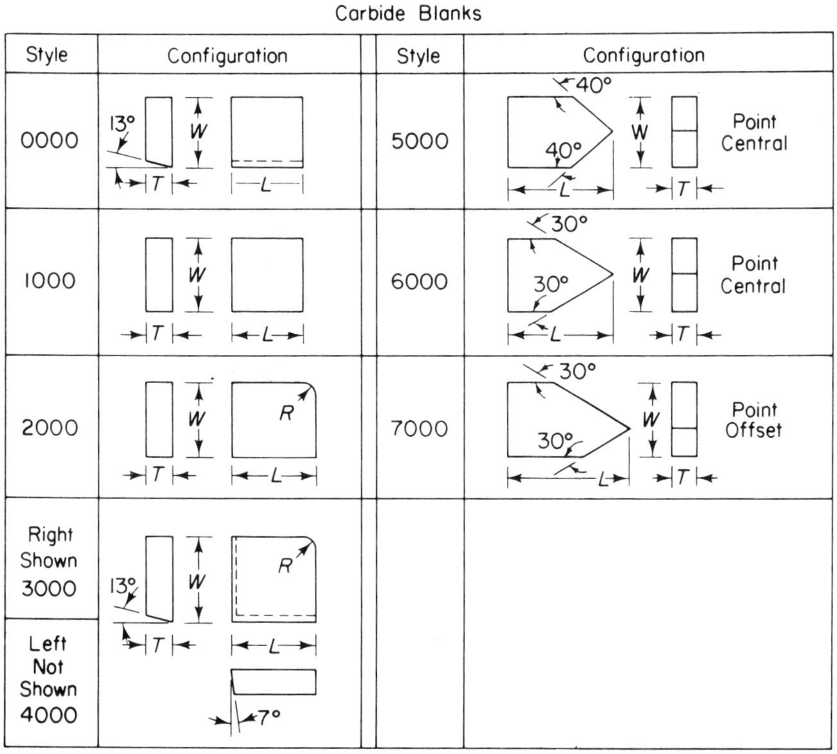

Sec. 4.3   Terminology, Clearance, Rakes and Chip Breakers   55

Figure 4.8

that sometimes this shearing action must be sacrificed in the interest of tool strength. When necessary, negative rake angles are used.

Figure 4.9(a) shows a tool with a 0° back rake, Fig. 4.9(b) one with a positive back rake, and Fig. 4.9(c) one with a negative back rake. A zero rake will project a line from the top of the tool through the center of the work, as shown. Positive rake sets up a "scooping" action. Negative rake sets up a "bulldozing" action. Tools may also have negative side rake, or combinations of positive and negative back and side rake.

Figure 4.9(b) shows the shearing force of the material acting on the tip of the tool. In Fig. 4.9(c) the shearing force is greater but acts back of the tip where the strength of the tool is greatest.

Chip breakers (Fig. 4.10) are used to control the directional flow of chips. This may not be necessary for ordinary lathe work but becomes important in automatic and semiautomatic boring, or any place where chip production could interfere with the cutting operation. Therefore, chip breakers are used to curl and break off chips into small

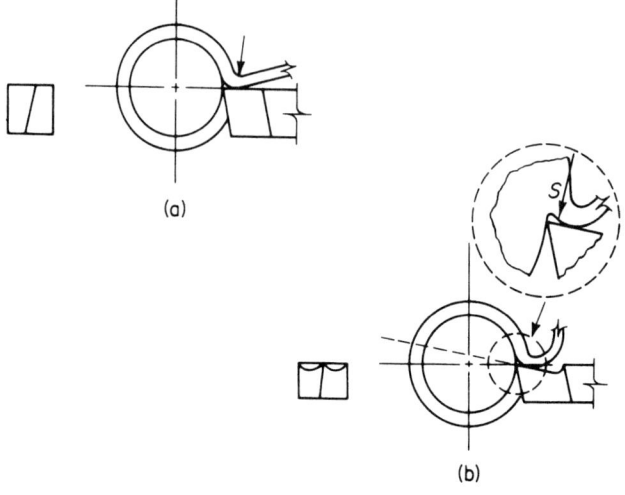
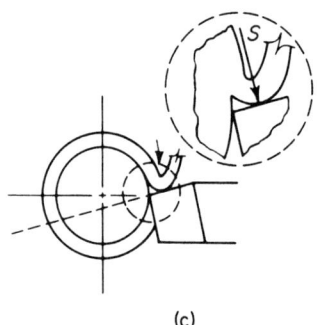

Figure 4.9

pieces so that they may be removed easily by coolants, or air, or simply allowing them to fall into the chip pan of the machine.

There are several varieties of chip breakers that may be used on any type of tool. In general, the groove needs to be slightly longer than the depth of the cut. The width of the groove, the depth of the groove, and the width of the land (or area) between the cutting edge and the groove depend on the feed used. Some typical chip breaker dimensions are shown in Fig. 4.10(a).

Figure 4.10(a) has a ground in groove. The land on this chip breaker may be up to 5° negative rake. Figure 4.10(b) and (c) show the step type of chip breakers. They may be parallel, or at an angle, to the cutting edge for better control of chip flow. The flat section at the bottom of the step-type chip breaker may itself create a positive, zero, or negative rake at the top of the tool. This is shown in the Fig. 4.10(b) inset, where the chip breaker has a positive rake ground parallel to the face of the tool.

Sec. 4.4   Milling Cutters 57

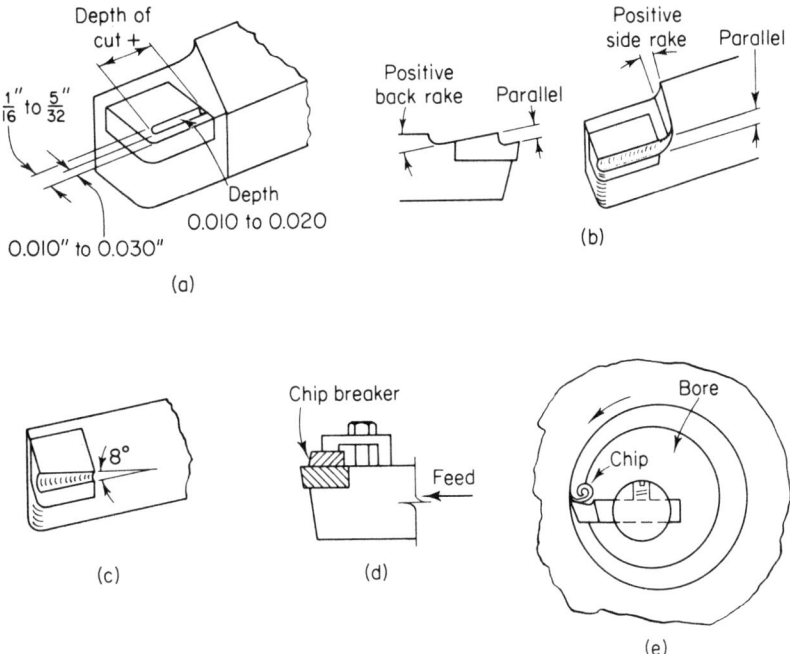

Figure 4.10

Figure 4.10(d) shows a mechanical chip breaker that can be set at any desired distance from the end of the tool edge. Figure 4.10(e) shows the advantage of having the chip breaker on a boring tool. The chip curls into a tight roll, breaks off, and falls out of the hole.

4.4 MILLING CUTTERS

Figure 4.11(a) shows a high-speed steel milling cutter. Relief and rake angles are applied to milling cutters in the same manner that they are applied to tool bits. The milling cutter is actually a multiple-tool-bit arrangement.

Figure 4.11(b) shows the relief and rake angles for a side milling cutter. This cutter has both axial and radial rake. The axial rake angle is the angle made by the peripheral cutting edge and a line parallel to the axis of the cutter. The radial rake angle is the angle made between the side cutting edge and the radius of the cutter. Rake angles of from 10 to 15° are used.

The "flat" created by the relief angle is called the *land*. The land is usually about $\frac{1}{32}$ to $\frac{1}{16}$ in. wide, depending on the diameter of the cutter. This is shown in Fig. 4.11(c). If the land becomes too

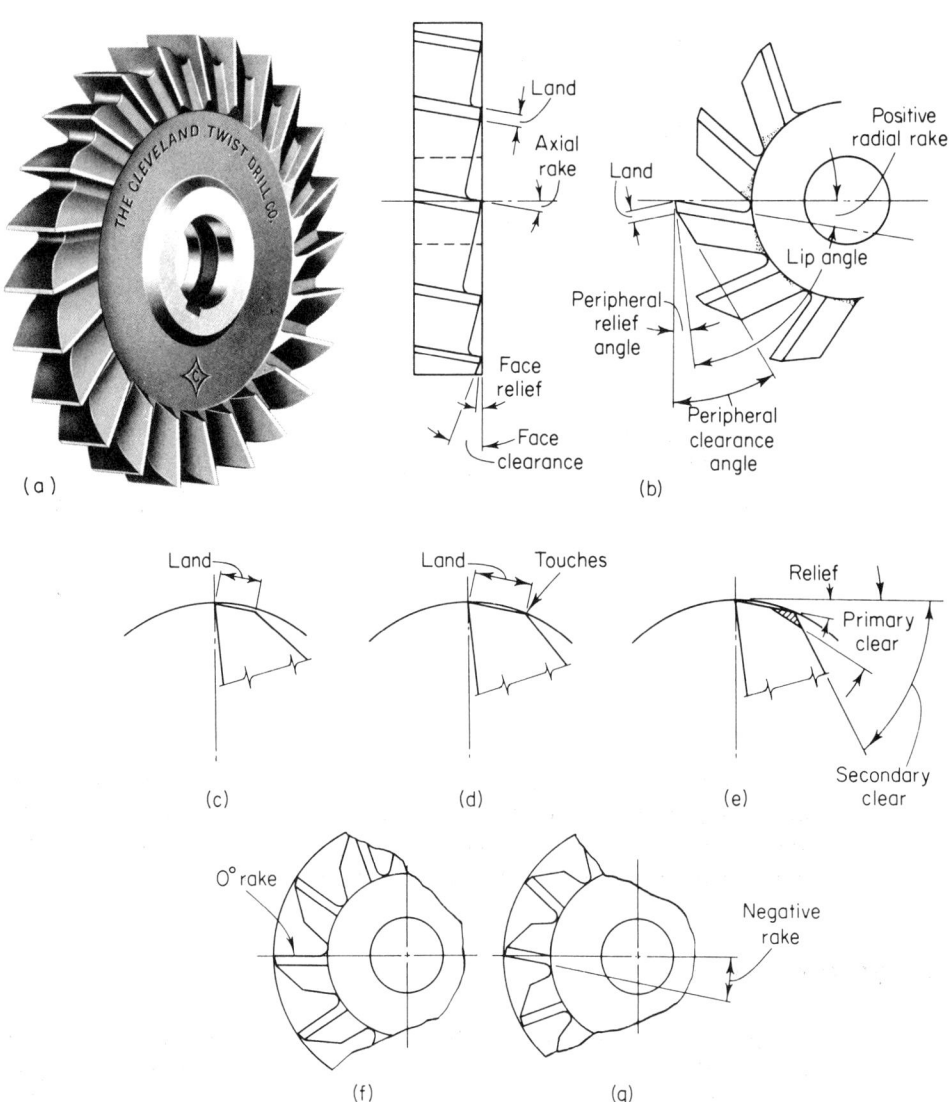

Figure 4.11

large, the heel of the land will touch the work and interfere with the cutting action, as shown in Fig. 4.11(d). This is avoided by grinding a relief angle and then getting the additional clearance by grinding a clearance angle. This operation is shown in Fig. 4.11(e). The primary clearance angle is usually ground double the size of the relief angle. This avoids the need for grinding the larger—usually 35°—clearance angle (secondary clearance).

Sec. 4.4   Miling Cutters   59

Relief angles are necessary to free the cutting edge. The size of the relief angle usually depends on the diameter of the cutter. In general, the relief angle should be about 3 to 5° and the clearance angle, as stated above, about 35°. The larger the diameter of the cutter, the smaller the relief angle needed. The relief and clearance angles should be kept as small as is consistent with the diameter of the cutter and the nature of the operation. Initially, the land should be small enough to achieve the desired results and still allow for several grindings of the cutter before the primary clearance is reground.

A milling cutter may have positive radial rake, as shown in Fig. 4.11(b), zero rake, as shown in Fig. 4.11(f), or negative radial rake, as shown in Fig. 4.11(g). Negative rake is used primarily for carbide milling cutters. A negative radial rake of -5 to $-10°$ in combination with $+5$ to $+10°$ positive axial rake is used for severe cutting operations such as interrupted cuts or for cutting tough steel.

In addition, cutters may be used to cut in various modes. Conventional, or up milling, is shown in Fig. 4.12(a), down milling in Fig. 4.12(b), and a combination of up and down milling in Fig. 4.12(c).

In conventional milling [Fig. 4.12(a)] the rotation of the cutter is creating forces that oppose the table movement. This requires a resistance force in the cutter that must overcome the thrust created by the

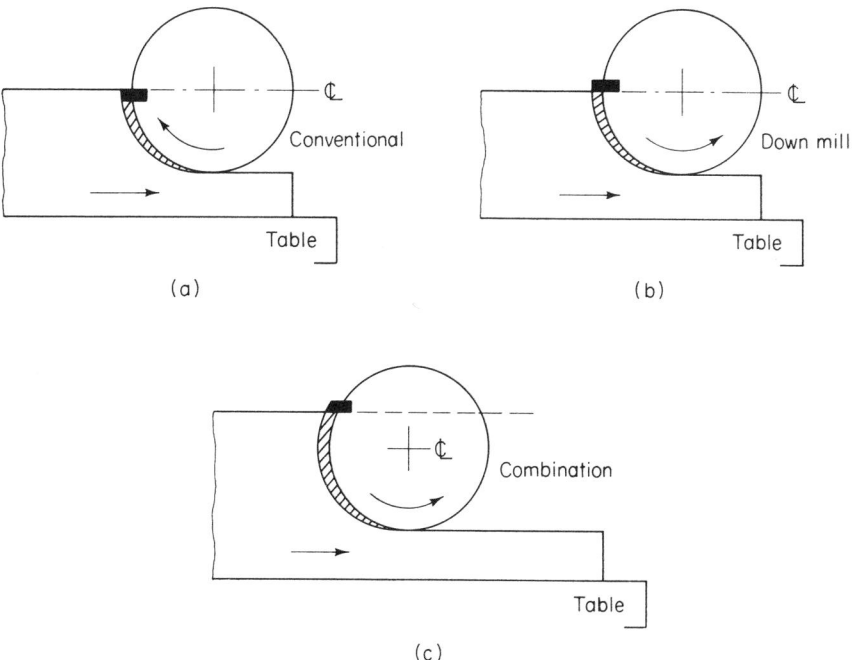

Figure 4.12

milling machine table. In down milling the cutter force created by its rotation acts in the same direction as the table thrust force. Thus less force is needed to cut through the material.

Another advantage of down milling over conventional milling is the fact that in down milling, chips welded to cutter teeth are dislodged as each tooth contacts the work. In conventional milling the chip is dragged into the work.

Also in conventional milling the tooth enters the work at zero depth and works its way into thicker cross sections of the work. In some materials this creates a work-hardening condition. In down milling the tooth enters the work at maximum depth. The thicker chip cross section dispenses the heat rapidly.

Still another advantage of down milling over up milling is the fact that in down milling, the force against the work is down against the table. In conventional milling the cutter is always trying to pull the work up off the table of the machine.

When cutting materials have a hard, or abrasive, surface, conventional milling allows the tooth of the cutter to come up underneath the crust and shear the hard chip off the work surface. Down milling will quickly dull the cutter tooth.

4.5 DRILLS AND REAMERS

Figure 4.13(a) shows a high-speed drill. The drill nomenclature is shown in Fig. 4.13(b). The relief angle for frills ranges from 8 to 12°. The chisel-edge angle is usually 135°. A standard ground drill has a

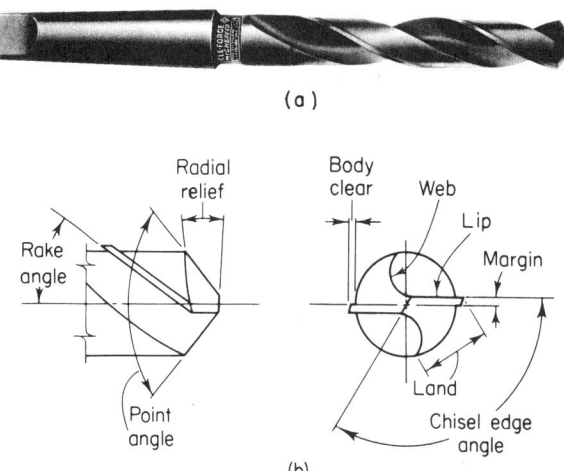

Figure 4.13

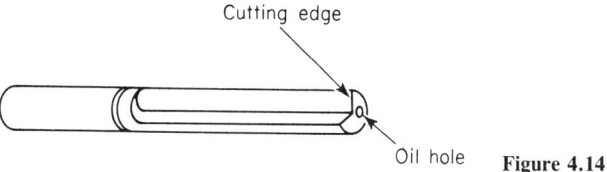

Figure 4.14

point angle of 118°. This angle is increased for harder materials and decreased for softer materials. Thus for hard, or tough, steels the point angle could be as much as 150°; for aluminum and cast iron, 60 to 130°; and for plastics and wood, 60 to 18°.

The rake angle is created by the helix of the flute. This is shown in Fig. 4.13(b). This angle may vary from 0 to 45°. The greater the helix angle, the less torque is required for the drilling operation. With steeper angles, the shearing effect is increased and so is the chip removal. Gun drills (Fig. 4.14) have 0° rake. The flute is parallel to the centerline of the drill.

QUESTIONS AND PROBLEMS

4.1. What are the ultimate purposes of any machining operation or sequence of operations?

4.2. Why is cutting tool geometry so important?

4.3. (a) Describe the mechanism by which a tool gets dull or fails.
(b) What methods and factors are used to overcome the fail mechanisms discussed in part (a)?

4.4. Describe the three types of chips discussed in this chapter.

4.5. What effect does a built-up edge have on:
(a) The cutting action of a tool?
(b) The surface finish on a workpiece?

4.6. What is cratering?

4.7. (a) Define and explain the term *orthogonal*.
(b) Define and explain the term *oblique*.
(c) Relate the two terms in parts (a) and (b) to the lathe cutting tool.

4.8. What effect does friction have on the operation of a tool bit? Discuss your answer.

4.9. What effect does heat have on a tool bit?

4.10. State the primary purpose of:
(a) A lubricant.
(b) A coolant.

4.11. List the recommended lubricants for:
(a) Steel and wrought iron.
(b) Brass and bronze.
(c) Cast iron.

4.12. (a) What is flank wear?
(b) What is its effect on the cutting operation?

4.13. (a) List the conditions associated with the selection of a tool material.
(b) Explain each condition and relate it to the efficiency of the cutting operation.

4.14. Which properties of a tool material must be present in order that the tool may operate efficiently?

4.15. List the six types of tool materials used in industry today. State the major advantages and disadvantages of each material.

4.16. State the optimum operating temperatures of each of the six tool materials used in industry today.

4.17. Describe the heat treatment of high-speed steel.

4.18. Make a drawing of a tool bit and label all the angles.

4.19. Define the following terms:
(a) Relief angle.
(b) Clearance angle.
(c) Rake angle.

4.20. (a) Draw a tool bit that has a 0° rake angle.
(b) Draw a tool bit that has a negative rake angle.
(c) Draw a tool bit that has a positive rake angle.
(d) Discuss the effect on chip formation of each of the tool bits in parts (a) through (c).

4.21. A tool bit has the following angles:

$$\text{front relief} = 8°$$
$$\text{negative back rake} = 4°$$
$$\text{negative side rake} = 7°$$
$$\text{end cutting-edge angle} = 10°$$
$$\text{side cutting-edge angle} = 6°$$

Find:
(a) The lip angle.
(b) The nose angle.
(c) The cutting angle.

4.22. (a) Make a drawing of a ground in chip breaker.
(b) What are the dimensional parameters of this type of chip breaker?

4.23. Make a drawing of a milling cutter tooth and label all parts.

Chap. 4 Questions and Problems

4.24. Make three drawings of a milling tooth, showing 0°, positive, and negative rake angles.

4.25. Describe:
(a) Conventional milling.
(b) Down milling.

4.26. Why does it take less force to down-mill than to mill conventionally?

4.27. What is the impact of "chip" welding on surface finish?   Explain your answer.

4.28. Discuss the "cutter tooth entrance" of a tooth into a workpiece in:
(a) Conventional milling.
(b) Down milling.

4.29. Make a drawing of a drill bit and label all parts.

5

Control Centers

5.1 THE CONTROL PANEL

There are many types of control panels manufactured by machine tool manufactures. Figure 5.1 and Fig. 5.2 show two panels that have the usual number of functions that the student will need. It should be understood that each manufacturer of computer-controlled machine tools uses a machine language to suit their particular needs.

Most systems have a self-test routine into the system that checks the various components in the system. Programs loaded into storage remain in storage until they are removed, or until the power to the machine is turned off.

Figure 5.1 shows the simplest type of control panel. It controls the X,Y motion of the milling machine table. Only one X,Y coordinate may be programmed at a time. The panel replaces the need to crank the table to a desired position. It permits coordinate designation either in the English, or metric system. It also allows absolute or incremental movements to be made.

5.2 MANUAL DATA INPUT

Manual Data input (MDI) is used to store information into the machine memory from a control panel as contrasted with putting data (program) into a post processor computer which then controls the machine tool.

Sec. 5.2   Manual Data Input   65

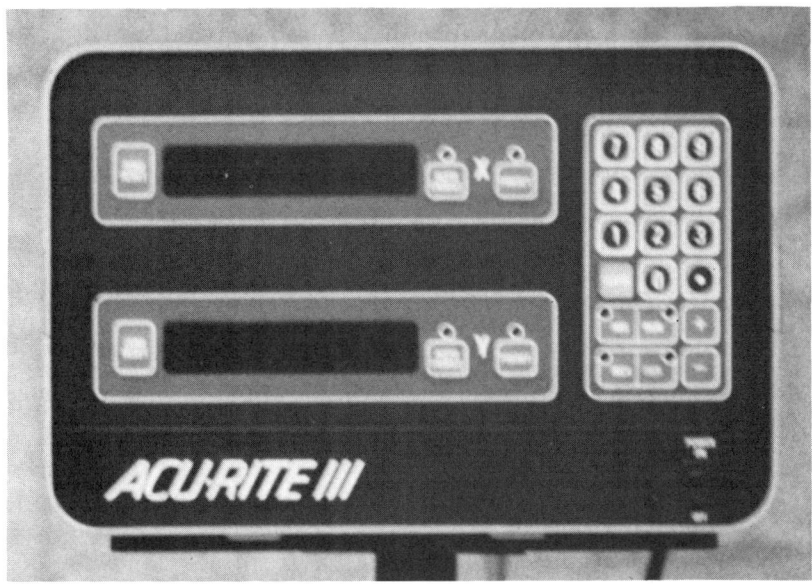

Figure 5.1   (with permission of Bridgeport Machine Tool Co.)

Figure 5.2   (with permission of Bridgeport Machine Tool Co.)

The control panel Fig. 5.2 allows for additional specific operations to be used. In this panel switches are used that permit various manual operations such as: Override, Modes of Operation, Functions, Axis motion and Abs/TLO.

The lower right part of the panel, Fig. 5.2, allows for X,Y and Z input in addition to various codes. The lower left part of the panel shows the readout of the Manual Data Input (MDI).

5.3 MANUAL CONTROL

Manual control buttons or switches provide the operator with the means to program, store and execute machine axis motions. When the manual mode is turned on, both the machine and its control mechanism are activated. In this mode, manual functions will be operational only when the control mechanisms and the machine tool are in manual mode.

Once the manual mode has been activated, the X, Y, Z values and the plus (+) and minus (−) directions are selected. In the manual mode the plus and minus motions are picked up from the last dimensional values programmed. They are not picked up from the zero datum point.

Thereafter, the type of motion to be used is selected for each axis. This may be a motion such as Step, or Jog. That means that in order to perform a "jog" the axis control must be in the manual mode of operation. Depressing the jog button will cause the table of the machine to move. Releasing the jog button will stop the motion of the table. In this system the speed at which the table moves is controlled by the feed rate override, or the design (O.E.M.).

By selecting the desired amount, incremental movements may also be made. By selecting the axis, direction and the distance to be moved, the table may be manually activated.

5.4 THE FLOATING ZERO

Many numerically controlled machine tools have built into their memory floating zero cabability. This means that at any time the machine tool may be directed to change its X,Y start point to a new location relative to the placement of the fixture. Once established, start points are rarely changed. However, should it become necessary to change the 0,0 start point to some place other than the initial point chosen, floating zero capability permits such a change.

The major advantage of a floating zero coordinate point is that it can be used to establish the 0,0 point of the cutter anywhere on the

Sec. 5.4   The Floating Zero   67

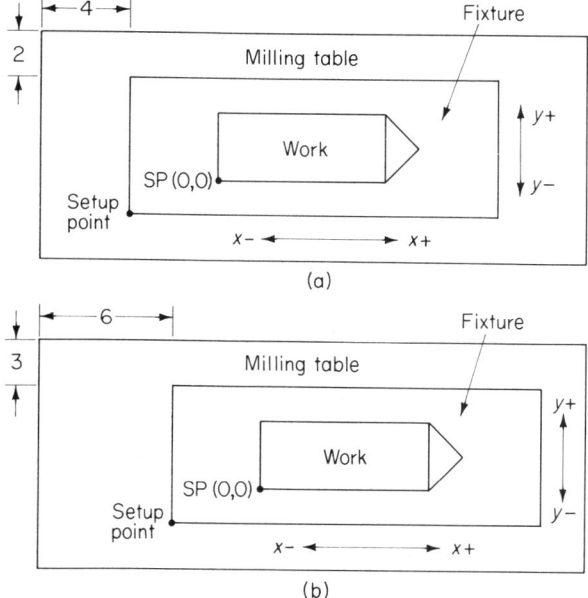

Figure 5.3

X,Y coordinates of the fixture position relative to the machine table. For example, a fixture may be strapped anywhere on a milling machine table. Thereafter, the cutter position is established in the X,Y direction relative to the new position of the fixture, or work. This procedure is much easier than first establishing the 0,0 point of the cutter and then locating the workpiece (or fixture) relative to the 0,0 point. It also permits the use of the fixture location point in subsequent setups.

Figure 5.3(a) shows a fixture mounted on a milling machine table. The setup point of the fixture is at X = 4, Y = 2. If it should be necessary, it may be mounted anywhere on the table. Once mounted, the 0,0 point is selected. It may be anywhere on, or off, the workpiece. Movement of the milling machine table is programmed relative to the selected 0,0 point. Thus in Fig. 5.3(b) the fixture is located at the coordinate point X = 6, Y = 3. The floating zero permits the start point of the workpiece to be shifted to its new location. There is no need to change the program. The same program may be used once the new zero position is established.

5.5 TOOL LENGTH OFFSET

As indicated earlier, manual modes are used for tool length offset (TLO) operations and for establishing and use of datum points. TLO is shown in Fig. 5.4. Once the TLO is established, other tools may be positioned in the Z direction relative to this Z zero position.

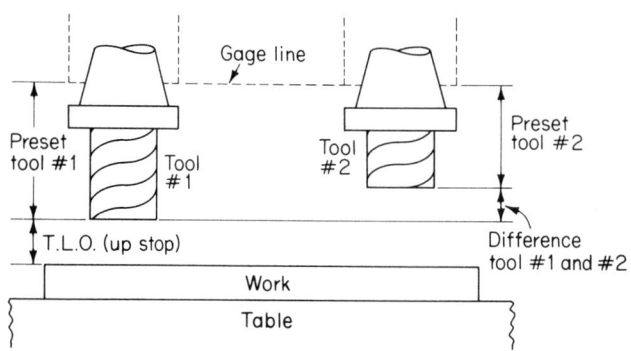

Figure 5.4

The procedure for setting the TLO for tools to be used is to select the longest tool, set its TLO, position the Z movement of the knee of the milling machine and secure it. Once the knee is set to Z-zero, all other tools may be set and Z-programmed to meet the needs of their own lengths. From Chapter 6 it will be seen that a tool change word (M6) must precede the coded tool word (ie, T3).

Once all offset values have been programmed, the usual procedure is to verify them relative to the TLO Z-zero position.

It should be noted that the TLO process can be applied when several cutters of different diameters are to be used. It may also be used with cutters that are oversize, undersize, or have been reground. Since the nominal size of the cutter has been programmed, the TLO process may be used to compensate for these variations in size.

Example 1.

Three milling cutters are to be offset to mill slots into a workpiece as shown in Fig. 5.5. Tool "A" is $4\frac{1}{4}$ inches long, tool B is 3 inches long and tool C is $3\frac{3}{4}$ inches long. Select an upstop TLO of 0.050 inches. Calculate the position of each tool in relation to the surface of the work.

Solution: Difference between A and B = 4.250 − 3.000 = 1.250 in.

$$1.250 + TLO = 1.250 + 0.050 = 1.300 \text{ in.}$$

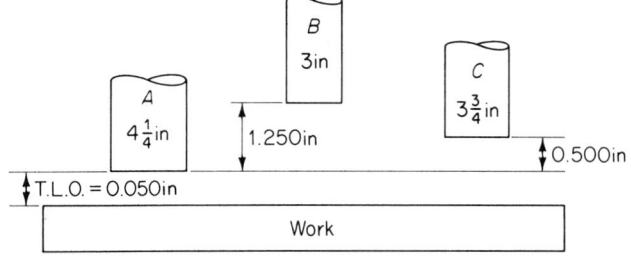

Figure 5.5

Difference between A and C = 4.250 − 3.750 = 0.500 in.

0.500 + TLO = 0.500 + 0.050 = 0.550 in.

5.6 FEED HOLD, CYCLE START, AND SINGLE BLOCK

The *free hold* command stops all axes feed motion in any mode of operation. This command does not stop the rotation of the spindle. To continue to perform according to the machine setup data (M.S.D.) of the program in process, the *free hold* button (switch) is pressed again. *Cycle start* may also reactivate the program.

Cycle start may be used to activate a program either (1) one line at a time; and (2) in the automatic mode, it will activate the entire program.

Single block allows the execution of one line of information. Once the machine is in the single block mode, pressing the cycle start button activates one line of information. To execute another line of information, the cycle start button must be pressed again.

5.7 EMERGENCY STOP

All CNC machine tools have some kind of emergency stop mechanism. A large bright red (yellow) mushroom-shaped button is usually located in full view of the operator for quick access. In case of an emergency (machine crash), pressing this button shuts down all power to the machine. The "power-up" procedure must be followed to reactivate the program.

QUESTIONS AND PROBLEMS

5.1. Research several machine tool companies and describe their control panels.

5.2. Identify the switches on the control panel in Prob. 5.1. Describe their purposes.

5.3. (a) What is the purpose of MDI?
(b) What is a "postprocessor"?

5.4. What first step must be taken before jogging the table of a milling machine?

5.5. Explain the purpose and use of the floating zero.

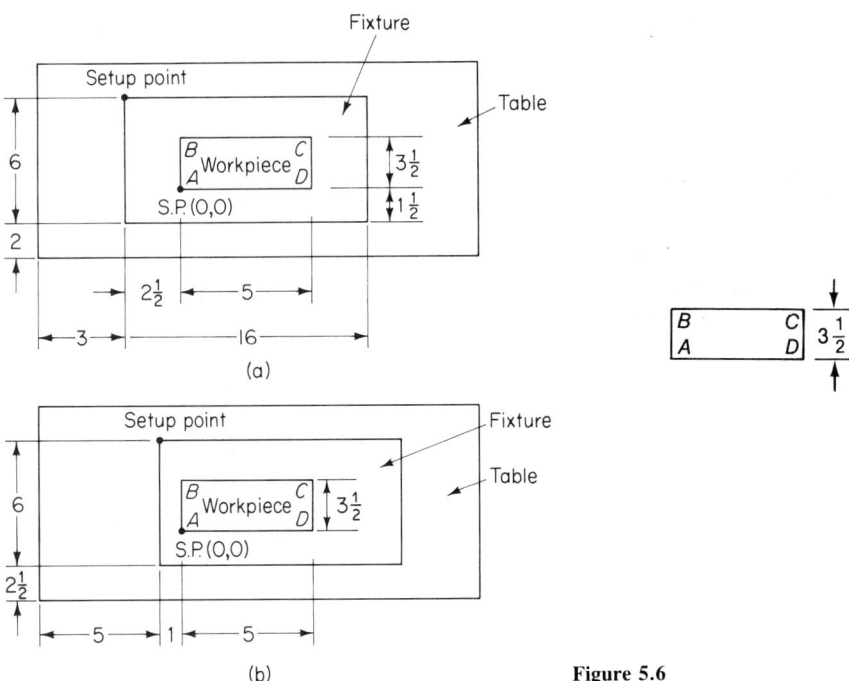

Figure 5.6

5.6. Figure 5.6(a) shows a fixture, and workpiece located relative to a milling machine table.
  (a) Calculate the point-to-point X and Y coordinates in Fig. 5.6(a) from the S.P.
  (b) Repeat Part (a) for Fig. 5.6(b).
  (c) Is there a difference?

5.7. Three milling cutters are to be offset to mill slots into the surface of block. Tool A is $3\frac{1}{4}$ in. long; tool B is 2 in. long, and tool C $2\frac{3}{8}$ in. long. Make a diagram similar to Fig. 5.2. Select the "up stop" TLO and show the position of each tool in relation to the surface of the workpiece.

5.8. Given a $2\frac{1}{4}$-in. "up stop" TLO for the following four end mills: $\frac{1}{4}$ in. diameter, $7\frac{1}{2}$ in. long; $\frac{3}{8}$ in. diameter, 3 in. long; $\frac{3}{4}$ in. diameter, 8 in. long; and $\frac{1}{2}$ in. diameter, 5 in. long. Show the position of all tools above the surface of the work.

5.9. Given a TLO "up stop" of 4 inches and a 1 in. diameter end mill 6 in. long. A second end mill is 4.375 in. long. Assume a grinding operation has removed 0.062 in. from the length of the second end mill. (a) How far above the work surface must the reground end mill be set? (b) Is the TLO distance in Part (a) the same as it was before grinding? Explain your answer.

Chap. 5 Questions and Problems

5.10. (a) Describe what is meant by the "free hold" command.
(b) How does it differ from the "cyclic start" command?

5.11. Is the "cycle start" command different from the "single block" command? Explain.

5.12. (a) Examine your CNC MDI panel. Does it have a method for activating an "emergency stop"? Identify it.
(b) How may a program be reactivated once an emergency stop has been activated?

6

Addresses, Codes, Blocks, and Lines

6.1 DEFINITIONS

A program to a computer memory is written in a format that is made up from a collection of lines of commands.   Such a line is

        N5 X2.5 Y1.25

By definition

    N is a code
    5 is an address

Together

    N5 is a block

A collection of blocks is a line:

    N5 X2.5 Y1.25

A line is written without spaces between blocks.   Thus

        N5X2.5Y1.25

A series of lines is a program.   A program sends a series of messages to the control mechanism of a computer that direct a machine tool to

Sec. 6.2   Line

TABLE 6.1 ALPHA-NUMERIC CODE

Code	Meaning
A	Angle designation
D	Depth of cut, canned cycles; cutter diameter compensation
F	Feed rate: in./min; in./rev
G	Preparatory function: motion command
H	Tool length offset
I	X-axis arc center offsets vector component direction, or parameter in canned cycles
J	Y-axis arc center offsets vector component direction, or parameter in canned cycles
K	Z-axis arc center offsets vector component direction, or parameter in canned cycles
L	Repeat counter in canned cycle, line number, and call
M	Miscellaneous function
N	Sequence number
O	Program number
P	Sequence number for canned cycle starts and/or jump to; also vector shift by degrees
R	Radius designation, arc cut; amount of cutter tool nose radius compensation; subprogram variables
S	Spindle speed, rev/min
T	Tool number or port number in tool changer
U	X-axis command, incremental
V	Y-axis command, incremental
W	Z-axis command, incremental
X	X-axis dimension command
Y	Y-axis dimension command
Z	Z-axis dimension command

Source: Courtesy of Bridgeport Machine Tool Co.

perform, or not to perform, required responses. It should be remembered that as advances are made in control mechanisms, new codes may be added, or old codes deleted, from existing lists such as those shown in Table 6.1. The purpose of this chapter is to familiarize the student with terminology and existing practices. Existing codes may be used as a foundation for dealing with new codes as they become available.

6.2 LINE

A line of information may consist of letters, numbers, or symbols (such as a period). Its purpose is to relay a message to the control system of a machine tool. It defines the parameter of a particular movement of a component of a machine. It may control the movement of a milling machine table, the spindle, the cutter, fluid flow, and so on.

Thus a line such as

N20 X1.25 Y-1.75

may direct a milling machine to move the table 1.250 in. in the positive X direction and a minus 1.750 in. in the negative direction. (These movements are explained more fully in later chapters.)

Lines of information are started with the code followed by a series of numbers. With each number change a new set of instructions is transmitted to the memory control mechanism of the machine tool.

The numbers associated with the N code are usually written in increments of 5 or 10. Should it become necessary, this permits the insertion of lines between two sequence numbers without disturbing the existing numerical order. In Example 1, the line starting with N12 is inserted into the program.

Example 1

Insert the line X1.62 Y1.47 into the program between lines N10 and N15.

```
N5    X2.5 Y1.25
N10              Z.5
N15              Z-.5
```

Solution

```
N5    X2.5 Y1.25
N10              Z.5
N12   X1.62 Y1.47    (inserted line)
N15              Z-.5
```

6.3 G CODES

A G code indicates a preparatory function. When used, it signifies a change in operation. It may direct the machine to change from a particular feed mode to a rapid mode; or it may direct an operation to take place in a clockwise, or counterclockwise direction, and so on. G codes are used to control numerous types of operations.

It should be noted that there are two types of G codes: modal and nonmodal. Modal G codes are retrained in memory until another G code from the same group is programmed to cancel it. A nonmodal

Sec. 6.4   X, Y, and Z Codes

TABLE 6.2 G CODES

Code	Meaning
G00	Rapid traverse
G01	Linear interpolation
G02	Circular interpolation, CW
G03	Circular interpolation, CCW
G17	Circular interpolation in the XY axes plane
G18	Circular interpolation in the XZ axes plane
G19	Circular interpolation in the YZ axes plane
G77	Facing cycle
G78	Pocket milling cycle
G79	Internal hole milling cycle
G80	Cancel canned cycle
G81	One drilling operation
G82	Spot drilling with a dwell
G83	Deep hole drilling with pecking
G84	Tapping cycle
G85	Boring cycle: feed down, dwell, feed up
G86	Boring cycle: feed down, spindle stop, rapid traverse return
G89	Boring cycle: feed in, dwell, feed out
G90	Absolute dimensioning mode
G91	Incremental dimensioning mode
G92	Preset register: zero shift

Source: Courtesy of Bridgeport Machine Tool Co.

G code is one that applies only to the line of information in which it appears. G codes will be explained as they are used in this book. Table 6.2 lists G codes that are useful when programming.

6.4 X, Y, AND Z CODES

X, Y, and Z are linear motion codes. They control the linear motion of the workpiece (milling machine) or tool (lathe). Thus, on the milling machine [Fig. 6.1(a)], the right-to-left, or left-to-right, movement of the table (workpiece) is programmed X. The front-to-back, or back-to-front, movement of the table is programmed Y. The movement of the tool (spindle) up or down is programmed Z. Each axis is programmed plus (+) or minus (−), depending on the direction of movement. The plus (+) sign may be omitted.

Figure 6.1(b) shows the motion of the tool bit when cutting a workpiece on a lathe. In this case the movement of the tool parallel to the centerline of the work is programmed Z. The movement of

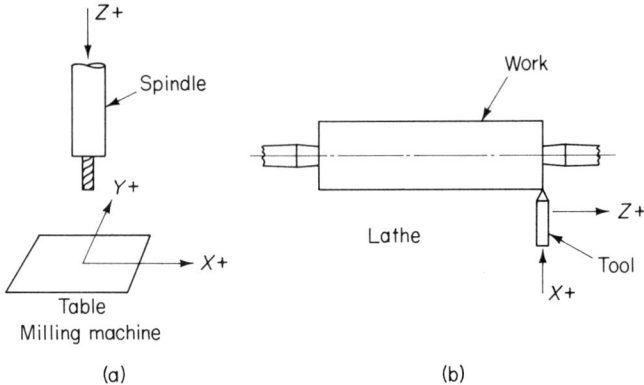

Figure 6.1

the tool perpendicular to the centerline of the work is programmed X. The digital value for the distance traveled may include the decimal point, Example 2; or it may be written without the decimal point, Example 3.

Example 2

        N50   X.500   Y.250

        N60   X.750   Y-1.000

Example 3

        N50   X00500   Y00250

        N60   X00750   Y-01000

Leading and trailing zeros may also be omitted with most machine tool memories. Example 4 shows the program when the zeros are omitted (Fig. 6.2).

Example 4

        N50   X.5   Y.25

        N60   X.75   Y-1.

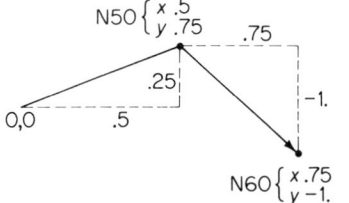

Figure 6.2

Sec. 6.5   I, J, and K Codes   77

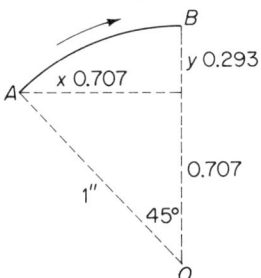

Figure 6.3

6.5 I, J, AND K CODES

I, J, and K are codes that define circular interpolation. Once the X, Y, and/or Z positions are established, the I, J, and K values are programmed. The latter codes are used when radii are to be machined.

In Fig. 6.3, assume that the center of the arc is at zero. The incremental movement from A to B in the X direction is 0.707 in. and in the Y direction it is 0.293 in. The program line for the arc movement from A to B in a clockwise incremental direction is

    X.707 Y.293 I-.707 J.707

Figure 6.4 shows the relationships of the I, J, and K values with

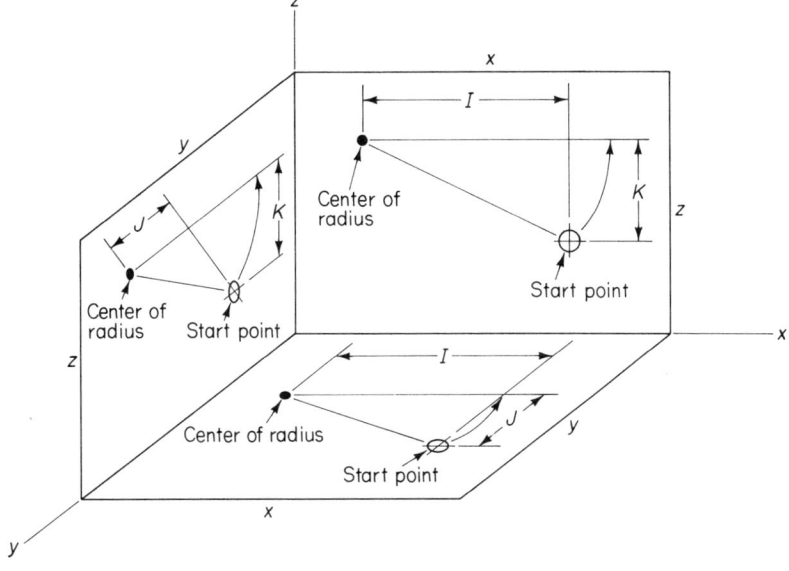

Figure 6.4

the X, Y, and Z directions, respectively. The plus or minus sign is used depending on the absolute or incremental mode of the operation to be performed.

6.6 S CODE

S is used as the code when spindle speeds are to be programmed. The rotation of the cutter on the milling machine, or the rotation of the workpiece on a lathe, is called the spindle speed. The unit is revolutions per minute. It is programmed with the letter S followed with a four-digit number.

Example 5

A spindle speed of 1000 rev/min is to be programmed. Write the block.

Solution

S1000

When coupled with an M03, M04, or M05, the spindle is directed to revolve.

Clockwise	M03
Counterclockwise	M04
Spindle Stop	M05

The zero will be included for clarity. The computer will disregard it.

6.7 F CODE

The F code is used to program the feed of a tool (lathe) or of a work table (milling machine). The feed is the rate of movement along a programmed path. The feed is usually programmed as inches per minute.

Example 6

Write the block for a feed of 13.3 in/min.

Solution

F0133

Feeds on a machine tool such as the milling machine may also be given in units of inches per revolution (in./rev) or inches per tooth (in./tooth). A feed that has units of inches per revolution may be

Sec. 6.8   T Code   79

converted to units of inches per minute by multiplying the former by revolutions per minute.

Example 7

Convert a feed of 0.015 in./rev to inches per minute when a cutter is operating at 400 rev/min.

Solution   To convert 0.015 in./rev to in./min, multiply as follows:

$$0.015 \frac{\text{in.}}{\text{rev}} \times 400 \frac{\text{rev}}{\text{min}} = 6.0 \frac{\text{in.}}{\text{min}}$$

A feed value in units of inches per tooth may be converted to inches per minute by converting inches per tooth to inches per revolution and then to inches per minute.

Example 8

Convert a feed value of 0.0015 in./tooth to units of inches per minute when a 16-tooth cutter is to be operated at 500 rev/min.

Solution   Multiply inches per tooth × teeth per revolution:

$$0.0015 \frac{\text{in.}}{\text{tooth}} \times 16 \frac{\text{teeth}}{\text{rev}} = 0.024 \frac{\text{in.}}{\text{rev}}$$

Then multiply inches per revolution by revolutions per minute:

$$0.024 \frac{\text{in.}}{\text{rev}} \times 500 \frac{\text{rev}}{\text{min}} = 12 \frac{\text{in.}}{\text{rev}}$$

6.8 T CODE

The T code is programmed to indicate a tool number in a magazine. Tools are placed in a particular position in this magazine and called up as needed. Programming a T and a number makes it possible to use that tool as many times as it is needed. When coupled with an M06 block, the tool change procedure is activated.

Example 9

Explain line N05 T10 M06.

Solution

N05	Line 5
T10	Tool number 10
M06	Activates the tool change

TABLE 6.3 M-CODE BLOCKS[a]

Code	Meaning
M00	Stop program
M01	Optional stop
M02	End of program and rewind
M03	Spindle on: CW
M04	Spindle on: CCW
M05	Spindle off
M06	Tool change
M07, 08	Coolant on
M09	Coolant off

[a] May vary with different manufacturer.

6.9 M CODE

These are function codes that are used to program special machine functions. Table 6.3 shows some of the more useful code.

6.10 D AND H CODES

The D and H codes are those that apply to preset tool length offset. This permits the programmer to use all tools as though they are all the same length. The code is used with either the tool change line or the spindle startup line. Once programmed it becomes the reference plane for all Z settings of all tools.

Example 10

Explain the tool change line

```
N05 T02 M06 H02
```

Solution

```
N05  Line number
T02  Calls for tool number 2
M06  Indicates that a tool change is needed
H02  Tool change offset reference applies to tool
     number 2
```

Example 11

Explain the tool change program

```
N05 T02 M06
N10 S1000 M03 H02
```

Solution

```
N05 T02 M06  Starting line
N10  Line number block
S1000 Spindle speed block
M03  Spindle is turned on clockwise
H02  Tool length offset block applies to tool num-
     ber 2
```

QUESTIONS AND PROBLEMS

6.1. Indicate in the format

```
N10X5.Y4.5Z1.2
N20X4.Y2.5Z-1.2
```

(a) the codes;
(b) the addresses;
(c) the blocks;
(d) the line;
(e) the program.

6.2. In Example 1 (Section 6.2), identify: (a) the codes; (b) the addresses; (c) the blocks; (d) the line.

6.3. From Table 6.1 select the code for: (a) tool length offset; (b) miscellaneous function; (c) X-axis arc; (d) Y-axis arc; (e) sequence number for canned cycle starts; (f) spindle speeds; (g) feed rate.

6.4. Describe the function performed by a "line" in a program.

6.5. Why is it advisable to number N-blocks in increments of five units? Explain and illustrate your answer.

6.6. How are G codes used?

6.7. Define and illustrate the concept of: (a) modal codes; (b) nonmodel codes.

6.8. Select the code necessary to perform: (a) the facing cycle; (b) the boring cycle: down, dwell, up; (c) the absolute dimensioning mode; (d) the incremental dimensioning mode; (e) linear interpolation; (f) rapid transverse.

6.9. Explain the motion actuated by: (a) a G2 and a G3 block; (b) an M3 and an M4 block.

6.10. Make a sketch of a milling setup. Indicate each of the three axis motions with a code and a plus or minus sign.

6.11. Repeat Prob. 6.10 for the lathe.

82   Addresses, Codes, Blocks, and Lines   Chap. 6

6.12. Write the program line for the following program when leading and trailing zeros are omitted.

```
N20  X+0.750  Y+1.250
N25                    Z-0.875
N30  X-0.750  Y-1.250  Z+0.875
```

6.13. Repeat Prob. 6.12 for the following program.

```
N10  X+2.255  Y+1.725  Z-0.075
N15                    Z+0.075
N20  X-2.255  Y-1.725
```

6.14. Explain the relationships between X, Y, Z directions and the I, J, K circular codes (Fig. 6.4).

6.15. Given a spindle speed of 50 rev/min and a feed of 25 in./min. Write the blocks.

6.16. Repeat Prob. 6.15 when the feed is 75.4 in./min and the spindle speed is 300 rev/min.

6.17. Given a spindle speed of 900 rev/min and a feed of 0.008 in./rev. Convert the feed to in./min.

6.18. Repeat Prob. 6.17 when the feed is 0.015 in./rev and the spindle speed is 600 rev/min.

6.19. Convert a feed of 0.002 in./tooth to units of inches per minutes when a 20-tooth cutter is operated at 600 rev/min.

6.20. Repeat Prob. 6.19 when an 18-tooth milling cutter is operated at a feed of 0.010 in./tooth and a spindle speed of 800 rev/min.

6.21. What block is needed to activate a T code?

6.22. Explain each block in the line

```
N20  T15 M06
```

6.23. Which performances do the following control: **(a)** M09; **(b)** T05; **(c)** H05; **(d)** M5?

6.24. Explain each of the blocks in the following line.

```
N05  T4 M06
N10  S900 M03 H04
```

7

Incremental CNC

7.1 COMMAND SYSTEMS

Commands are processed from a 1-in. tape (NC), directly from a teletypewriter, or from a computer (CNC). The system that controls the movements that results in the machining operations are essentially open, semiclosed, or closed loop.

The command lines, *in the form of blocks of information*, are either picked off a tape, or stored and released by the minicomputer. As noted later in this chapter, lines of commands are forwarded to servomotors in "blocks" of information. These signals control the movement of the table or spindle along the X, Y, or Z axes in three-axis machines. Two-axis machines have only the X and Y axes controlled. Four- and five-axis machines have one or two rotary degrees of freedom controlled as well as X-, Y-, and Z-axis control.

In *open-loop* control systems [Fig. 7.1(a)] the command is fed into the machine but no backfeed takes place. There is no check system built into the machine to determine whether or not the command has been carried out. Thus, in Fig. 7.1(a), the operator control console is used to set the initial conditions of the zero position, start the machine, and so on. The tape reader picks off the discrete bits of information from the tape and sends them to the distribution control unit, which directs the information to the servo control unit for a particular axis control. The command is then transferred to the servomotor, which carries out the command. In Fig. 7.1(a) the X and

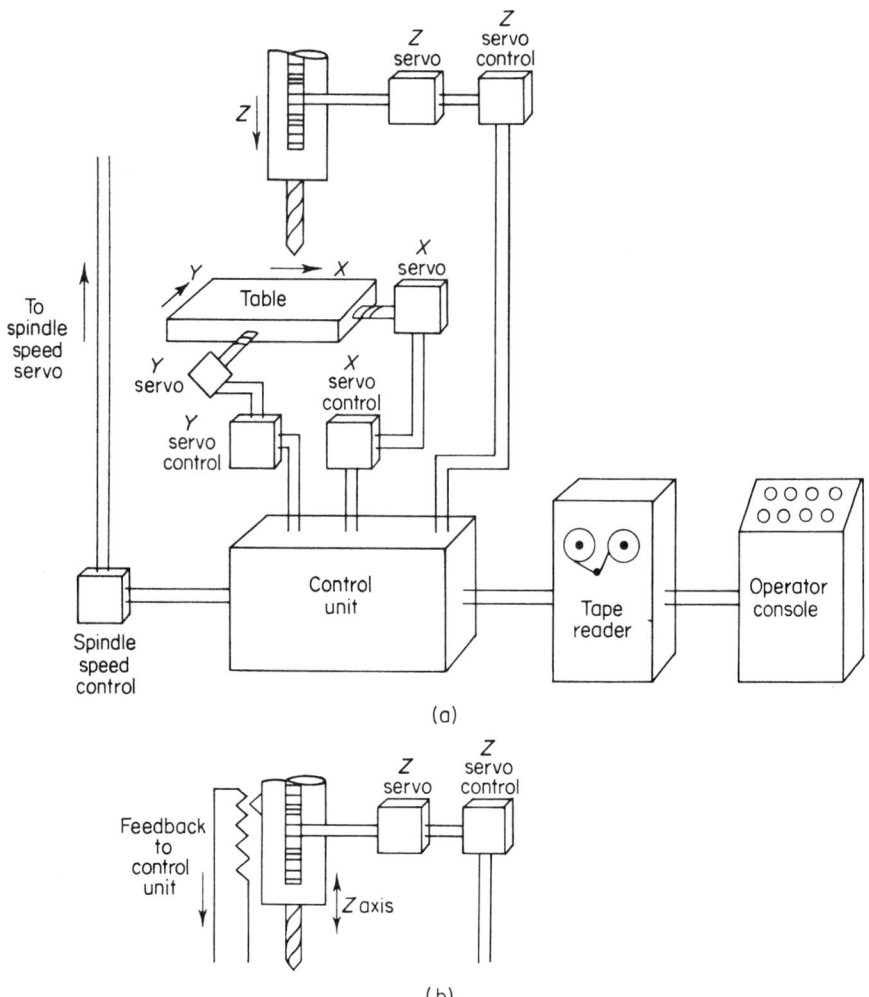

Figure 7.1

Y axes are controlled. Also included is a Z-axis control and a spindle-speed control. This is a three-axis system. It is to be noted that there is no feedback system to check on the command while it is being carried out.

Figure 7.1(b) shows a closed-loop system that has a transducer attached to each axis. The transducer converts the motion, as directed, from linear (or rotary) motion to an electrical impulse. This impulse is fed back into the control unit as a check on the input signal. If the input signal and the feedback signal match, the servo stops and the desired position of the tool is obtained. If the input signal and the feedback signal do not match, the error signal, the difference between the input and feedback signals, takes over as the input signal. It remains the control signal until a match is achieved. Thus the correction is made, and the operation continues.

The transducer may be either digital or analog. The digital signal is a pulse signal proportional to the unit motion of the tool or worktable. This may be accomplished by counting the number of pulses needed for a given movement, or by matching pulses to the input signal.

The analog signal is continuous and proportional to the continuous movement of the table, rotation of the spindle, or movement of the spindle. As in the case of a digital transducer, the continuous input voltage is matched to the transducer output voltage to achieve accurate positioning.

Semiclosed-loop systems vary from closed-loop systems in that the latter verify and control the actual position of the tool with reference to the work. In contrast, the semiclosed-loop system verifies the position of the work relative to the tool in an *indirect way*. It may measure the position of the work by counting the number of turns of a lead screw, or by driving a gear and comparing the number of turns against a standard. Obviously, this is not as accurate as a closed-loop system. It is, however, less expensive to manufacture.

There are many different "languages" that have been adapted, or developed, by machine tool manufacturers to serve their own special purposes. APT I, II, and III are word languages with which the programmer describes the parts as they appear on the engineering drawing. Programmers have at their command a 107-word language that they use when describing the movement of the tool or work. This description is fed into a computer that translates the directions to a tape, thus saving much calculation time. ADAPT is a simpler version of APT. AUTOSPOT, AUTOPROMPT, and SNAP are but a few of the additional word languages available. They are all used to save time and length of tape required in point-to-point or contour control. As indicated, the simplest type of command is the point-to-point system.

If we assume a two-axis machine (X and Y), the movement of the table may be controlled so that it will move along the X or Y axis. One company recommends visualizing the movement of the cutter. Thus, in Fig. 7.2(a), when the *cutter* machines from A to B, *away from* the operator, the direction is plus (+). When the *cutter* moves *toward* the operator, the direction is minus (−). The movement of the *cutter* from *right to left* [Fig. 7.2(b)] (C to D) is plus; from *right to left*, the X direction is minus.

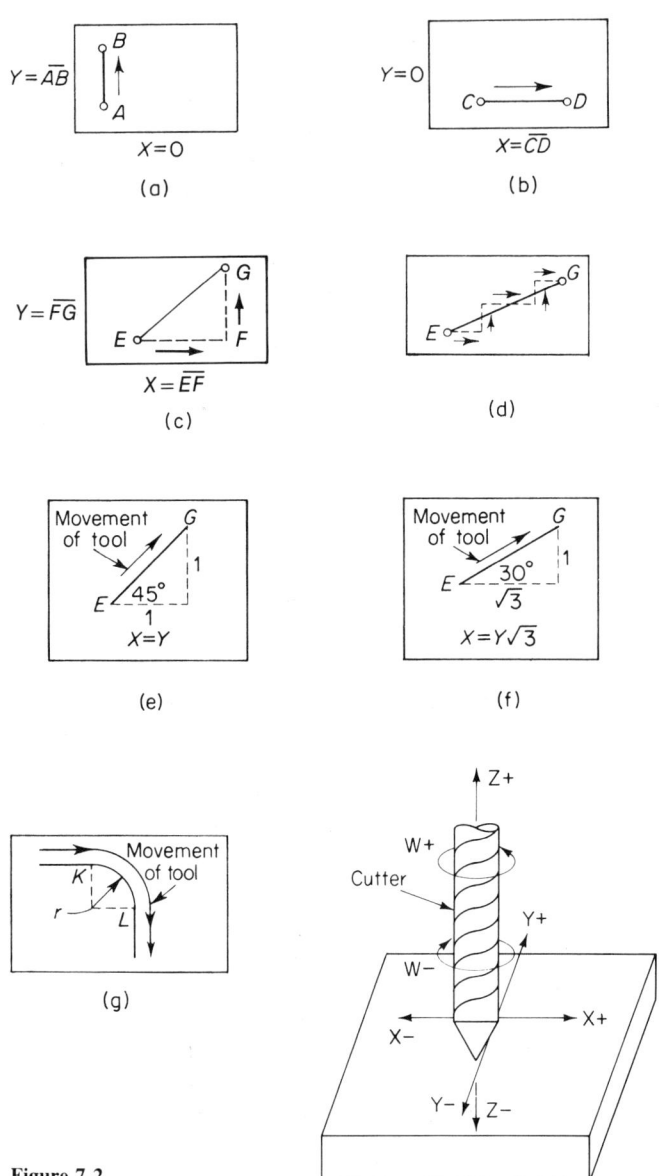

Figure 7.2

TABLE 7.1 CUTTER MOVEMENT

(+) Plus Direction	(-) Minus Direction
X to right	X to left
Y to rear	Y to front
Z up	Z down
W clockwise	W counterclockwise

Sec. 7.1 Command Systems

It is to be noted that the up movement of the spindle is plus (there are systems that reverse the + and − directions of Z. In Fig. 7.2(c) the movement of the tool is positive (E to F) along the X axis and positive (F to G) along the Y axis. Commands are incorporated in a block of information that results in the movement of the tool from E to G. These movements are shown in Fig. 7.2(h) and Table 7.1.

If line EG is to be machined, the X-Y movements are reduced so that the movement of the tool approximates the straight-line movement EG. This is shown in Fig. 7.2(d). It should be noted that the greater the number of X-Y movements, the more nearly does the tool approach the straight line EG. This is one of the disadvantages of the point-to-point system. However, the greater the number of XY movements, the greater the accuracy achieved. In contour machining, the required number of points needed may become excessive. Despite this, many straight-line operations may be done with the point-to-point system.

In continuous-path systems the tool follows the desired shape since the commands are far more descriptive for point-to-point systems. They are capable of causing the table to move so that X is some function of Y, designated X = f(Y). Thus a machine table may be directed to move so that X always equals Y [Fig. 7.2(e)], or so that $X = Y\sqrt{3}$ [Fig. 7.2(f)], or along the arc KL of radius r, as shown in Fig. 7.2(g). In the last instance the end points of the arc KL and the equation of the curve are needed. The direction of movement of the table (or tool) is controlled by the use of + and − notations. The movement of the tool may be programmed into the tape, or machine, so that it generates the desired curve.

It should be evident that the more complex the contour, the more complicated is the mathematics needed. Thus for continuous-path numerical control, a computer is usually needed to calculate the command signals. It is also possible to establish command signals with the use of very accurate drawings. This method may be more expensive and not as accurate or as fast as a computer. In sophisticated setups, the computer is inserted in the loop in either the open-loop or the closed-loop system.

7.2 CODES

Two types of number codes are generally used. The *straight binary system* is one used for continuous control in which multiple simultaneous motions may be programmed in one column. It produces lengths of tape that are shorter than the binary-coded-decimal system used in NC programming.

The *binary-coded-decimal system* is based on the standards established by the Electronics Industries Association (EIA). The tape is 1.000 in. wide, has eight channels available, and has a sprocket-feed channel as shown in Fig. 7.3(a). The punched holes and the channels are 0.100 in. apart. One hole is called a *bit*. A series of holes punched in a row across the tape is called a *code*. A *word* is a complete set of characters comprising a complete unit of information.

In the EIA system the sum of the number of holes in a code must be *odd*. This is referred to as an *odd-parity check*. If a particular code does not have an odd number of holes punched in a row on the tape, the reader will not function. The *American Standard Code for*

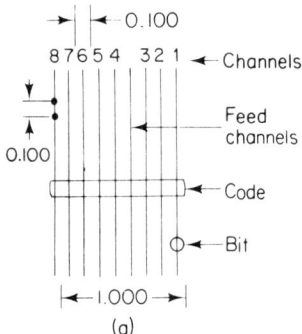

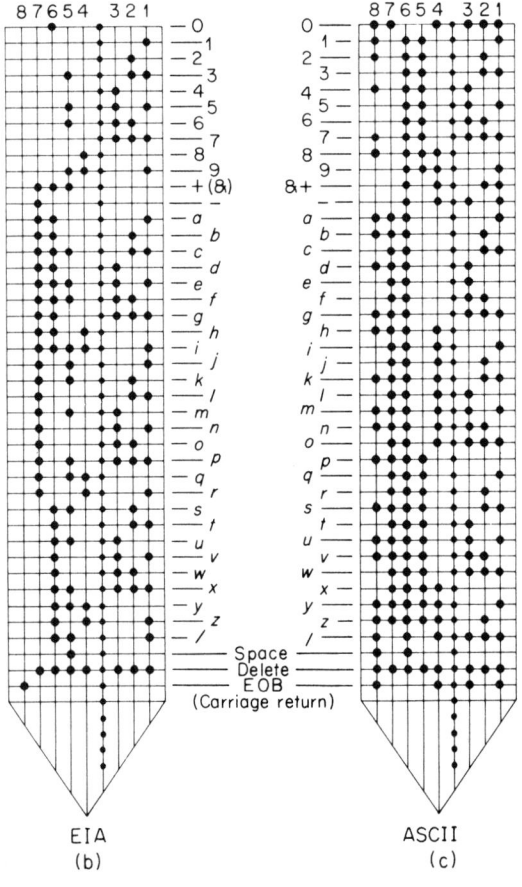

Figure 7.3

Sec. 7.2   Codes

Information Interchange, known as ASCII, uses the even-parity check. The EIA system is shown in Fig. 7.3(b) and the ASCII system in Fig. 7.3(c).

It should be noted that tape readers have the capability of reading holes either mechanically or electronically. In the mechanical principle a sprocket tooth enters a punched hole in the tape, causing an electrical contact, or combination of contacts, to be made. If a hole is missing in a particular row of holes, the tooth will not enter the tape and electrical contact is not made.

In the electronic system, a photoelectric reader uses a light source on one side of the tape and a row of solid-state photoelectric cells lined up on the other side of the tape. Figure 7.4(a) shows one such photoelectric reader. As indicated earlier, the information is received and stored until a block of information has been completed. The *end of block* (EOB) carriage return signals the release of this information.

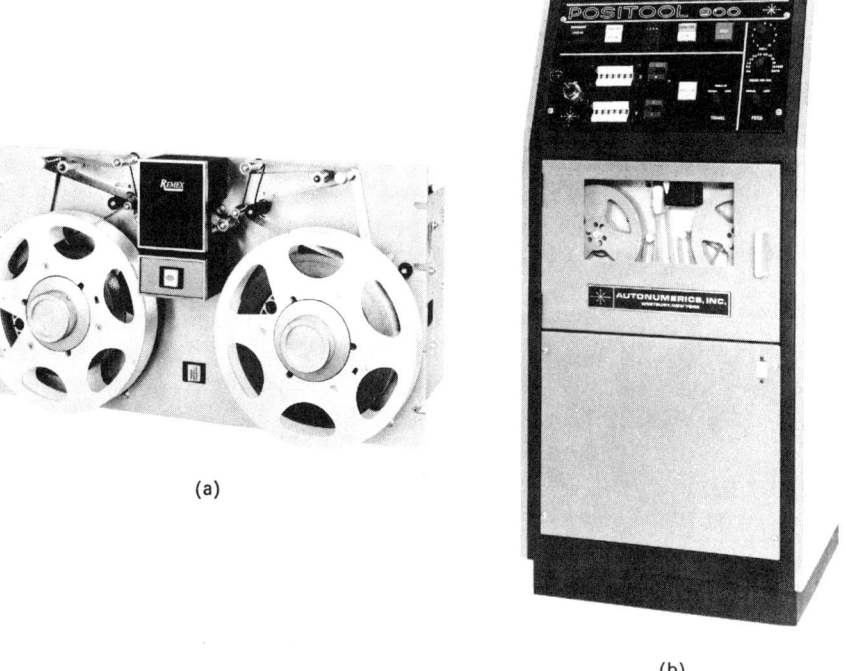

Figure 7.4 [(a) Courtesy of Ex-Cell-O Corportation, Remex Division; (b) courtesy of Autonumerics, Inc.]

TABLE 7.2 WORD ADDRESS CODES

Code	Meaning
N	Sequence number
G	Preparatory function, followed by preparatory function code number
X	X-axis movement
Y	Y-axis movement
Z	Z-axis movement
I	Circular interpolation, X-axis arc
J	Circular interpolation, Y-axis arc
K	Circular interpolation, Z-axis arc
F	Feed rate, followed by a feed code number
S	Spindle speed, followed by a speed code number
T	Tool, followed by a tool code number
R	Radius, followed by a radius code number
A	Angle
M	Miscellaneous, followed by a function number
+	Cutter direction
-	Cutter direction
EOB	Carriage return

This makes it possible for several movements to take place at the same time. Thus it is possible for a tool to travel directly from E to G in Fig. 7.2(c) without first moving to F along the X axis and then along the Y axis to G. A tape reader console is shown in Fig. 7.4(b).

Codes are developed from a combination of letters and numbers. The letter addresses a particular register. Thus an N25 means the twenty-fifth line, station address, or sequence number. Additional addresses are shown in Table 7.2 and Section 6.1. The codes produced on the tape are shown in Fig. 7.3.

It is important that all commands related to a block be typed without a space between commands. Thus Fig. 7.5(a) shows a block of commands, Fig. 7.5(b) the typed equivalent, Fig. 7.5(c) the same line with the zeros omitted, and Fig. 7.5(d) the tape. Note that the end of block (EOB) is accomplished using the carriage return. Table 7.3 lists some of the commands available to the programmer.

Programs should start with a *begin* code followed by a carriage return (EOB). At the end of the program, an end code and carriage return will cause the tape to rewind. Upon reaching the B code, the rewind stops and all machine dials should have returned to zero (or start). The machine and program are ready to repeat the operation.

The *feed command* uses a feed rate code of F followed by four digits. The decimal point is presumed to be between the last two

Sec. 7.2 Codes

G	F	X	Y	Z	I	J
08	0200	+00420	−01650		+00000	+01650

(a)

G08F0200X 00420Y-01650I 00000J 01650

(b)

G08F0200X.42Y−1.65I0.0J1.65

(c)

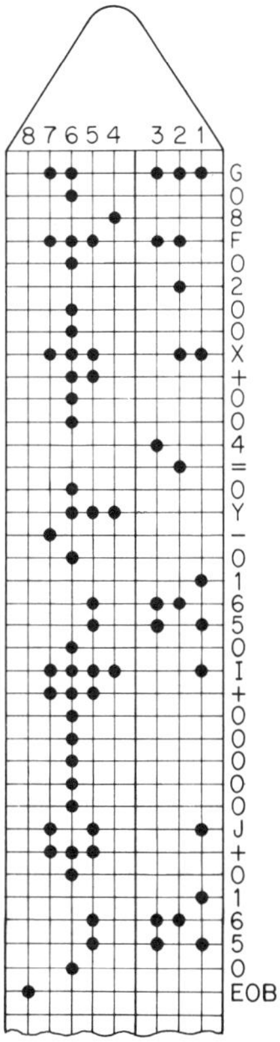

(d)

Figure 7.5

digits. Thus

$$F000\wedge0$$

If it is desired to feed at 125.7 in./min, the command is written as

$$125.7 \text{ in./min} = \text{F125}\wedge7$$

If it is desired to feed at 62.5 in./min, the command is written as

$$62.5 \text{ in./min} = 0625$$

The *spindle speed* rate is written in hundredths. Thus

$$700 \text{ rev/min} = \text{S07}$$

TABLE 7.3 AVAILABLE COMMANDS

Command	Operation
G00	Rapid traverse
G01	Linear feed
G02	Clockwise arc
G03	Counterclockwise arc
M03	Turn spindle on clockwise
M04	Turns spindle on counterclockwise
M05	Turn spindle off
M06	Tool change
M07	Turns coolant on
M09	Turns coolant off
S00	Spindle speed in hundredths
T00	Tool change number

An F9999 code is a rapid command. In this system all feed commands should precede the linear commands. Thus

```
N30    F9999    X6.0    Y-2.0
```

In other systems the feed rate follows the linear commands. Thus

```
N30    X6.0    Y-2.0    F6.0
```

Note: Leading and trailing zeros may be omitted. If a system does not need leading and trailing zeros, and they are used, the system will disregard them. A single zero will be used after the decimal point in this text for clarity. It may be omitted.

7.3 PROGRAM

There are two systems used to develop the coordinates of a point: the incremental and the absolute. The *incremental system* references all dimensions *point to point*. It always uses the *previous point as the zero start point* for the next movement. The *absolute system* references all dimensions from an *initially* determined fixed start point.

Consider Fig. 7.6(a) as redimensioned in Fig. 7.6(b). It has been redimensioned *incrementally*. The dimensions should be shown at each point. Note the omission of the + sign. (The use, or nonuse, of the plus sign is optional.)

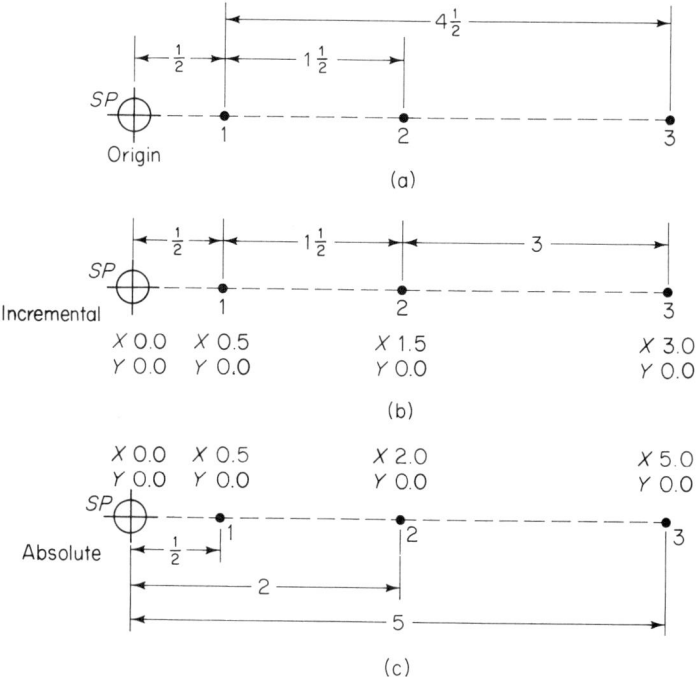

Figure 7.6

The incremental analysis is shown in Table 7.4. Note that the sum of the X and Y movements add to zero. The absolute system is developed in Chapter 9. For the present, Fig. 7.6(a) is shown redimensioned in the absolute mode in Fig. 7.6(c). The absolute analysis is shown in Table 7.5. Note the X = 0, Y = 0 return to SP (⊕).

TABLE 7.4 INCREMENTAL ANALYSIS

Move	X	Y
SP	0.000	0.000
⊕ to 1	0.500	0.000
1 to 2	1.500	0.000
2 to 3	3.000	0.000
3 to ⊕	-5.000	0.000
Sum	0.000	0.000

TABLE 7.5 ABSOLUTE ANALYSIS

Move	X	Y
SP	0.000	0.000
to 1	0.500	0.000
1 to 2	2.000	0.000
2 to 3	5.000	0.000
3 to	0.000	0.000

7.4 CUTTER POSITION: X AND Y INCREMENTAL MOVEMENTS

In this section we deal with the positioning and the movement of the cutter (table) from one point to another and back to the start point. Positioning commands are shown as five-digit numbers. Thus 3.465 in. reads

03465

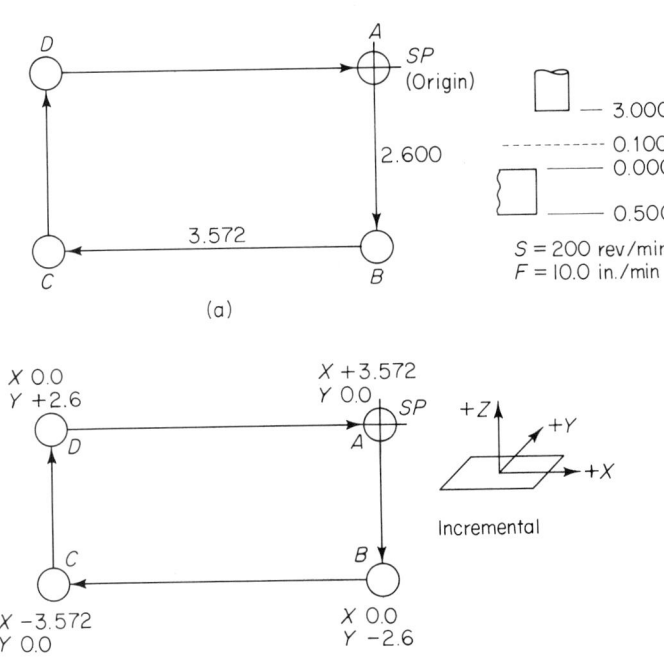

Figure 7.7

Sec. 7.5 Linear Contouring

TABLE 7.6(a)

Move	X	Y
A to B		-02600
B to C	-03572	
C to D		02600
D to A	03572	
Sum	00000	00000

TABLE 7.6(b)[a]

N Address	G Code[b]	X Axis	Y Axis	M Function	Comment
N10	G00[a]	0.0[b]	-2.6		A to B
N20		-3.572	0.0		B to C
N30		0.0	2.6		C to D
N40		3.572	0.0		D to A
N50				M02	(end of program)

[a]Zeros following the decimal point may be omitted.
[b]G00, rapid traverse.

or, omitting the zeros, it reads

3.465

0.247 in. reads

00247

or, omitting the zeros, it reads

.247

The decimal point is between the third and fourth digits from the right. Figure 7.7(a) show the points to be machined. Figure 7.7(b) shows the incremental dimensions inserted into the drawing. It is advisable to insert these dimensions into the drawing before writing the program.

The analysis for Fig. 7.7(a) is as shown in Table 7.6(a). The program would be as shown in Table 7.6(b).

7.5 LINEAR CONTOURING

In Fig. 7.8(a) the cutter should start at A, proceed to B, and then to C. It proceeds from A to B generating a square movement of

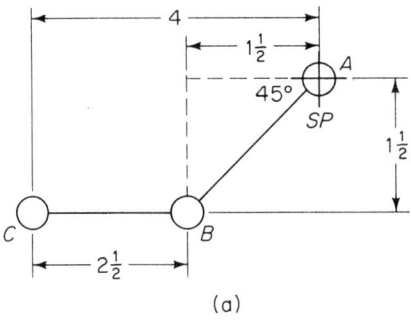

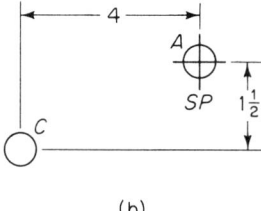

Figure 7.8

$1\frac{1}{2} \times 1\frac{1}{2}$ in. The cutter traces the 45° as long as both dimensions are programmed on one line. It then traces the distance from B to C.

Machines capable of linear interpolation, or contouring, make the movement from A to C directly. Thus the program [Fig. 7.8(b)]

X-4.0   Y-1.5

will cause the cutter to move directly from A to C.

Example 1

(a) Write the incremental analysis for the equilateral triangle in Fig. 7.9(a).
(b) Assume a spindle speed of 350 rev/min and a feed rate of $\frac{1}{4}$ in. Complete the program for Fig. 7.9(b).

Solution The *origin* is a point from which all dimensions are taken. The *start point* (SP) is a point from which the cutter starts. Figure 7.9(b) has been redimensioned for linear XY movement. It is important to note that the origin has been coded 0,0.

(a) The analysis is shown in Table 7.7(a).
(b) The program is shown in Table 7.7(b).

Sec. 7.5 Linear Contouring

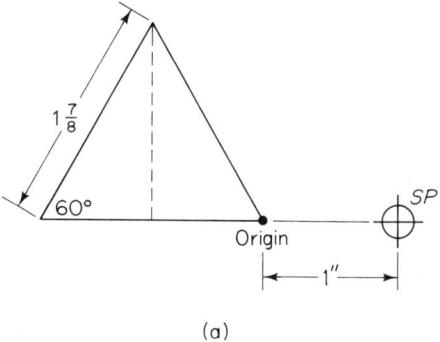

(a)

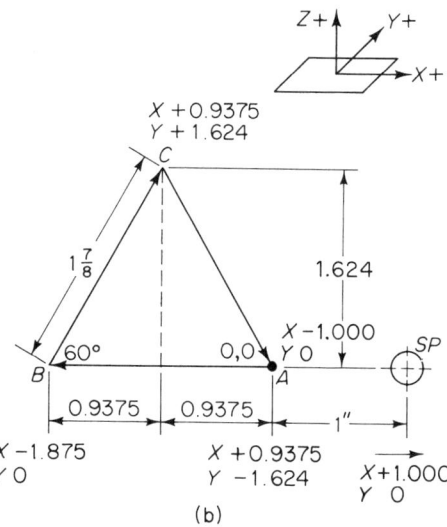

(b)

Figure 7.9

Example 2

(a) Analyze Fig. 7.10(a) in the incremental mode.

(b) Assume a spindle speed of 325 rev/min and a feed rate of 3.8 in. Complete the program for Fig. 7.10(b).

Solution (a) The analysis for Fig. 7.10(a) is shown in Table 7.8(a). The X and Y dimensions have been inserted in Fig. 7.10(b).

(b) The program is shown in Table 7.8(b).

TABLE 7.7(a)

Move	X	Y
⊕ to A	-1.000	0.000
A to B	-1.875	0.000
B to C	0.9375	1.624
C to D	0.9375	-1.624
A to ⊕	1.000	0.000

TABLE 7.7(b)[a]

N Address	G Code[a]	X Axis	Y Axis	F in/min	S ft/min	M Function	Comment
N10					S350	M03	Spindle CW
N20	G00[b]	-1.0	0.0				SP to A
N30	G01[b]	-1.875	0.0	F25			A to B
N40		0.9375	1.624				B to C
N50		0.9375	-1.624				C to A
N60	G00	1.0	0.0				A to SP
N70						M02	(end of program)

[a]Zeros following the decimal point may be omitted.
[b]G00, rapid traverse; G01, linear feed.

Sec. 7.6   Z Movement

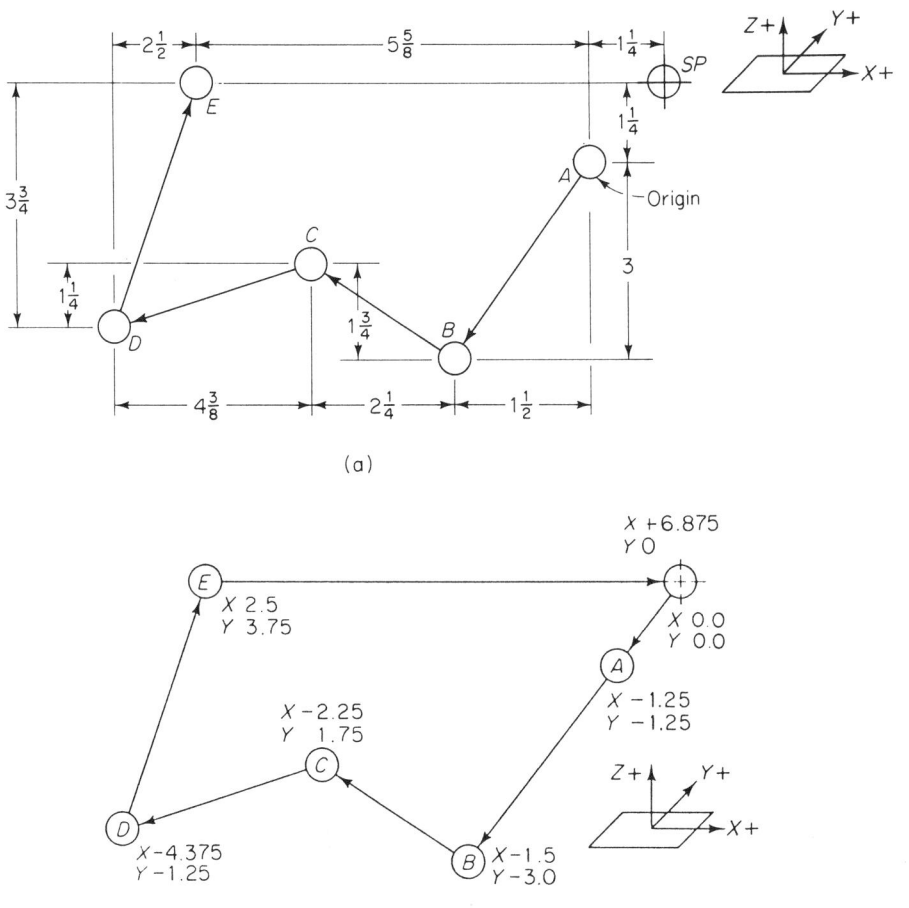

Figure 7.10

7.6 Z MOVEMENT

The vertical movement of the quill is accomplished by programming plus or minus Z commands. The plus (+) direction is up. (The plus direction should be checked in your service manual.) In Fig. 7.11(a) the start point and origin are at ⊕. All movements are taken from ⊕.

Example 3

(a) Analyze the X, Y, Z movements [Fig. 7.11(a)] in the incremental mode and tabulate your results.

TABLE 7.8(a)

Move	X	Y
SP	0.000	0.000
SP to A	-1.250	-1.250
A to B	-1.500	-3.000
B to C	-2.250	1.750
C to D	-4.375	-1.250
D to E	2.500	3.750
E to SP	6.875	0.000

TABLE 7.8(b)

N Address	G Code	X Axis	Y Axis	F IPM	S FPM	M Function	Comment
N10					S325	M03	Spindle CW
N20	G00	-1.25	-1.25				SP to A
N30	G01	-1.5	-3.0	F3.8			A to B
N40		-2.25	1.75				B to C
N50		-4.375	-1.250				C to D
N60		2.5	3.75				D to E
N70		6.875	0.0				E to SP
N80						M02	(end of program)

G00, rapid traverse.
G01, linear feed.

Sec. 7.6 Z Movement

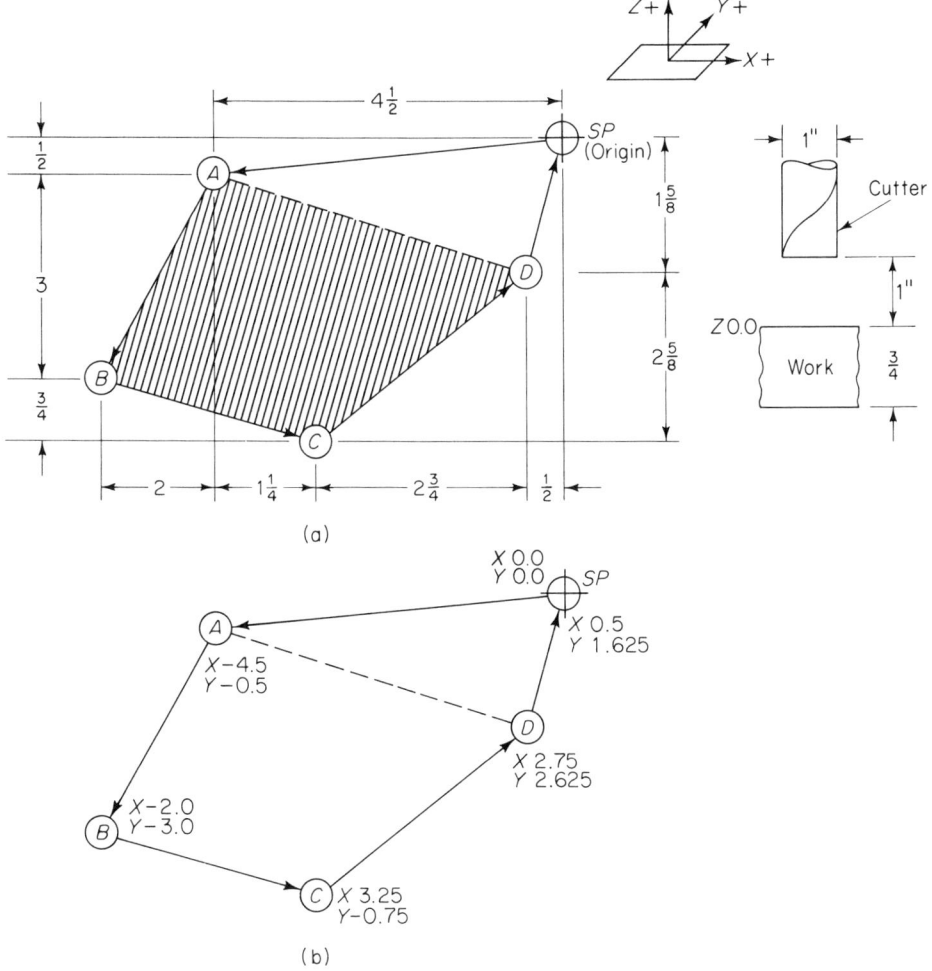

Figure 7.11

(b) Complete the program given a feed of 1.5 in./min and a spindle speed of 300 rev/min.

Solution (a) In Fig. 7.11(b) the X, Y, Z movements in the incremental mode are as shown in Table 7.9(a). It should be noted that the Z movements are programmed on a separate line. Later it will be seen that the Z movements may be combined with either the X or the Y movements.

(b) The complete program is shown in Table 7.9(b).

TABLE 7.9(a)

Move	Command	X	Y	Z
SP to A	5	-04500	-00500	-01000
At A	10			
A to B	15	-02000	-03000	
B to C	20	03250	-00750	
C to D	25	02750	02625	
At D	30			
D to SP	35	00500	01625	01000
Sum		00000	00000	00000

TABLE 7.9(b)

N Address	G Code[a]	X Axis	Y Axis	Z Axis	F IPM	S FPM	M Function	Comment
N10							M03	Spindle CW
N20	G00	-4.5	-0.5					SP to A
N30						S300		At A
N40	G01			-1.0	F1.5			
N50		-2.0	-3.0					A to B
N60		3.25	-0.75					B to C
N70		2.75	2.625					C to D
N80	G00			1.0				Z axis to 1" above Z0
N90		5.0	1.625				M02	D to SP (end of program)

[a]G00, rapid traverse; G01, linear feed.

Sec. 7.6   Z Movement   103

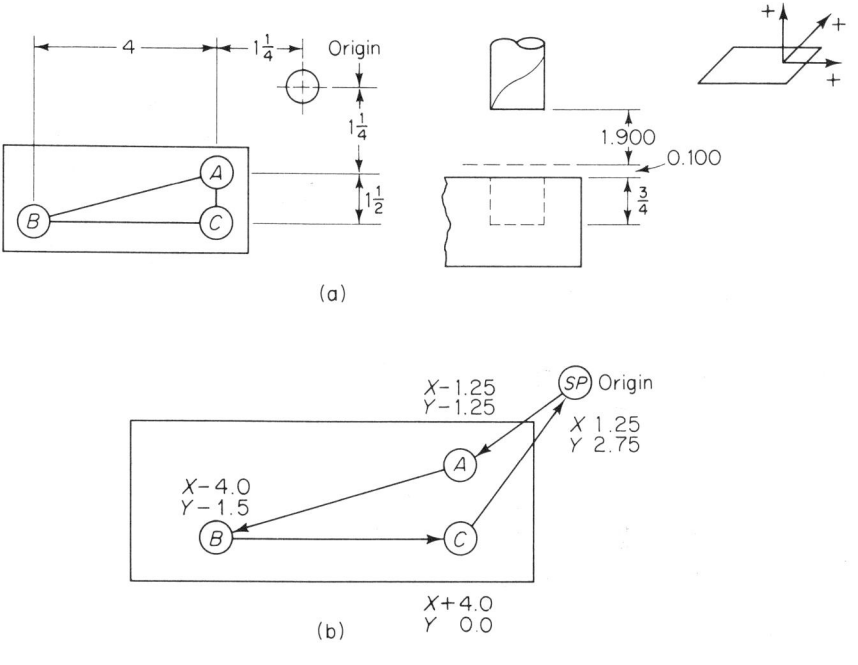

Figure 7.12

Example 4

(a) Do the analysis for Fig. 7.12(a) using the data that follow.

(b) Program the movement of the table for Fig. 7.12(a). Assume a two-flute end mill. The distance from Z zero to the bottom of the tool is 2 in. Use a rapid feed rate for a distance of 1.900 in. and a feed rate of 15 in./min for the remaining approach of 0.100 in. The feed rate, while cutting, is 15 in./min. Drill each hole $\frac{3}{4}$ in. deep. The spindle speed is 950 rev/min.

Solution   (a) The analysis for Fig. 7.12(a) is shown in Table 7.10(a).
(b) The program is shown in Table 7.10(b).

Example 5

Figure 7.13(a) is to be programmed in the incremental mode.

(a) Rework the dimensions in the incremental mode.
(b) Make a new drawing and insert the dimensions.
(c) Write the program when the feed is 2.750 in./min and the spindle speed is 350 rev/min.

TABLE 7.10(a)

Move	Command	X	Y	Z
SP to A	20	-01250	-01250	-01900
0.100 above work	30			-00850
0.750 into work	40			00850
0.100 above work	50			
A to B	60	-04000	-01500	-00850
0.750 into work	70			00850
0.100 above work	80			
B to C	90	04000	00000	-00850
0.750 into work	100			02750
2.000 above work	110			
C to SP	120	01250	02750	

TABLE 7.10(b)

N Address	G Code[a]	X Axis	Y Axis	Z Axis	F IPM	S FPM	M Function	Comment
N10						S950	M03	Spindle CW
N20	G00	-1.25	-1.25					SP to A
N30				-1.9				Z axis is 0.100 above work
N40	G01			-0.850	F15			Z machined 0.750 into work
N50	G00			0.850				Z axis 0.100 above work
N60		-4.0	-1.5					A to B
N70	G01			-0.850	F15			Repeats line N40
N80	G00			0.850				Returns to 0.100
N90		4.0	0.0					B to C
N100	G01			-0.850	F15			Repeats N40
N110	G00			2.750				Returns 2 in. above work
N120		1.25	2.75					C to SP
N130							M02	(end of program)

[a]G00, rapid traverse; linear feed.

Sec. 7.6   Z Movement   **105**

Solution (a) The reworked dimensions are as follows [Fig. 7.13(b)]:
B to C:

$$\tan 20° = \frac{0.750}{x}$$

$$x = \frac{0.750}{\tan 20°} = 2.061$$

$$y = 0.750$$

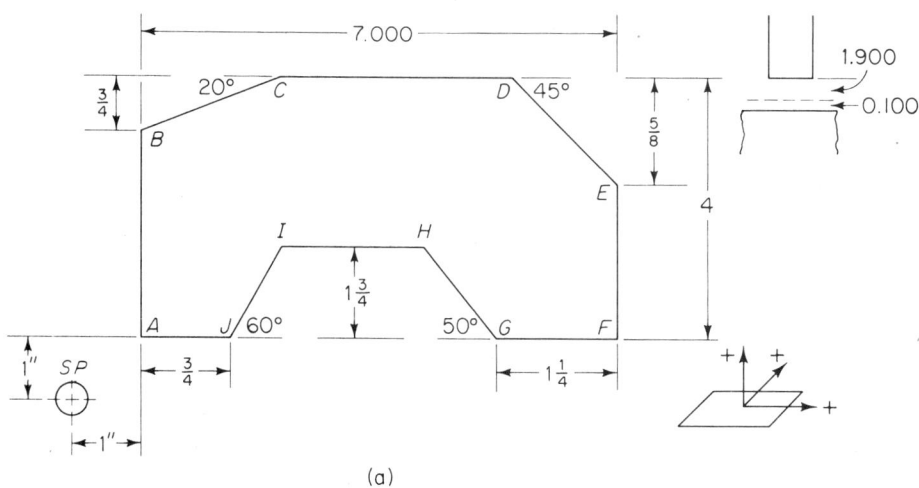

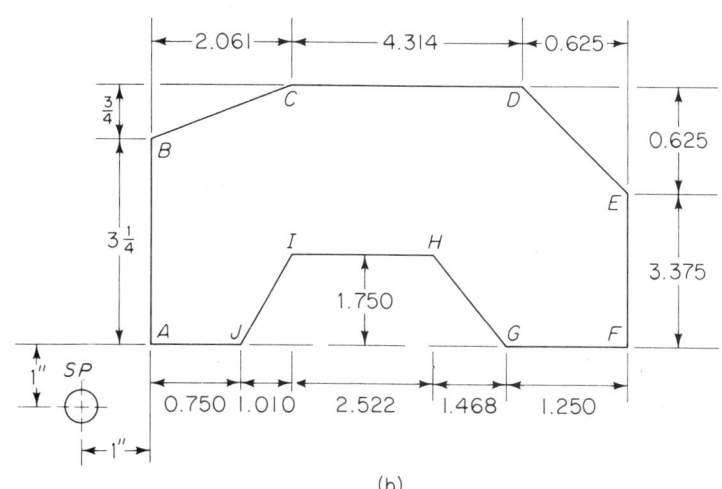

Figure 7.13

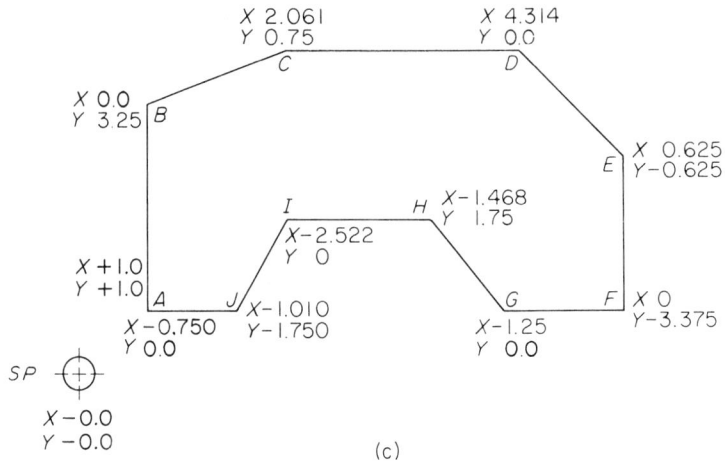

Figure 7.13 (Continued)

The coordinates are

        X02061   Y00750

D to E: The coordinates are

        X00625   Y-00625

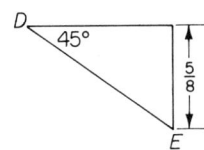

G to H:

$$\tan 50° = \frac{1.750}{x}$$

$$x = \frac{1.750}{\tan 50°} = 1.468$$

$$y = 1.750$$

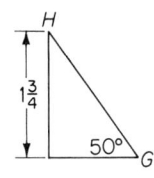

The coordinates are

        X-01468   Y01750

I to J:

$$\tan 60° = \frac{1.750}{x}$$

$$x = \frac{1.750}{\tan 60°} = 1.010$$

$$y = 1.750$$

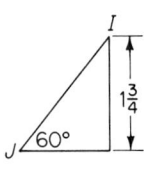

Sec. 7.6 Z Movement

The coordinates are:

$$X-01010 \quad Y-01750$$

The coordinates at each point are [Fig. 7.13(c)] as follows:

SP to A:
X = 1.000
Y = 1.000

A to B:
X = 0.000
Y = (4.000 − 0.750)
  = 3.250

B to C:
X = 2.061
Y = 0.750

C to D:
X = (7.000 − 0.625 − 2.061)
  = 4.314
Y = 0.000

D to E:
X = 0.625
Y = −0.625

E to F:
X = 0.000
Y = (4.000 − 0.625)
  = −3.375

F to G:
X = −1.250
Y = 0.000

G to H:
X = −1.468
Y = 1.750

H to I:
X = (7.000 − 1.468 − 1.010
  − 1.250 − 0.750)
  = −2.522
Y = 0.000

I to J:
X = −1.010
Y = −1.750

J to A:
X = −0.750
Y = 0.000

A to SP:
X = −1.000
y = −1.000

(b) These dimensions have been inserted into Fig. 7.13(c).

(c) The analysis for Fig. 7.13(a) is as shown in Table 7.11(a) and inserted in Fig. 7.13(b). The program for Fig. 7.13(c) is shown in Table 7.11(b).

TABLE 7.11(a)

Move	Address	Code	X	Y	Z	F
At SP	N5					
SP to A	N10	B	1.000	1.000	-1.900	Rapid
At A	N15				-0.100	2.75
At A	N20					
A to B	N25		0.000	3.250		
B to C	N30		2.061	0.750		
C to D	N35		4.314	0.000		
D to E	N40		0.625	-0.625		
E to F	N45		0.000	-3.375		
F to G	N50		-1.250	0.000		
G to H	N55		-1.468	1.750		
H to I	N60		-2.522	0.000		
I to J	N65		-1.010	-1.750		
J to A	N70		-0.750	0.000		
At A	N75					
A to SP	N80		-1.000	-1.000	2.000	Rapid
At SP	N85	R				

TABLE 7.11(b)

N Address	G Code[a]	X Axis	Y Axis	Z Axis	Spindle Speed (rev/min)	Feed (in./min)	M Function	Comment
N10	G00	1.0	1.0		350		M03	Spindle CW
N20				-1.9				SP to A
N30				-0.1				At A
N40	G01					2.75		At A
N50			3.250					A to B
N60		2.061	0.75					B to C
N70		4.314	0.0					C to D
N80		0.625	-0.625					D to E
N90			-3.375					E to F
N100		-1.25	0.0					F to G
N110		-1.468	1.75					G to H
N120		-2.522	0.0					H to I
N130		-1.01	-1.75					I to J
N140		-0.75	0.0					J to A
N150	G00							At A
N160		-1.0	-1.0	2.0				A to SP
N170							M02	(end of program)

[a]G00, rapid traverse; G01, linear feed.

QUESTIONS AND PROBLEMS

7.1. What is the purpose of the floating zero in a numerical control setup?

7.2. Explain how a numerical control machine checks itself when it is used to machine several similar workpieces.

7.3. Draw and explain the purpose of each unit in the block diagram for an open-loop numerical control system. Why is it called an open-loop system?

7.4. Draw and explain the purpose of each unit in the block diagram for a closed-loop numerical control system. Why is it called a closed-loop system?

7.5. Describe the purpose and operation of a transducer in numerical control.

7.6. How does a semiclosed-loop system differ from a closed-loop system? Illustrate.

7.7. Select one of the word languages mentioned in Section 1.1 and describe its purpose.

7.8. Which axes are usually controlled on:
(a) A two-axis machine?
(b) A three-axis machine?
(c) A four-axis machine?
(d) A five-axis machine?

7.9. Define and explain a digital signal and an analog signal.

7.10. (a) Explain the principle of point-to-point command for actuating the transverse movement of the table of a machine.
(b) Do as in part (a) for the longitudinal movement of the table.

7.11. Assume that two holes are to be drilled at 30° to the longitudinal traverse on a numerical control drill press. In using the point-to-point command system, what is the movement of the table?

7.12. Assume that a slot is to be machined at 30° to the horizontal traverse on a numerical control milling machine. How does the cutter move in a point-to-point command system?

7.13. Why is it difficult to machine a complicated contour with the point-to-point command system? Illustrate the movement of the tool when cutting a simple radius contour with a point-to-point system.

7.14. Explain why it is possible to cut contours, or slots, at an angle by using the continuous-path command system. Describe the principle.

7.15. What is the purpose of using a computer in numerical control?

7.16. What is the difference between EIA and ASCII binary-coded-decimal systems?

7.17. (a) Program the numbers 1 through 9 using columns 1, 2, 3, and 4 in Fig. 7.3(b).
(b) Repeat part (a) for Fig. 7.3(c).

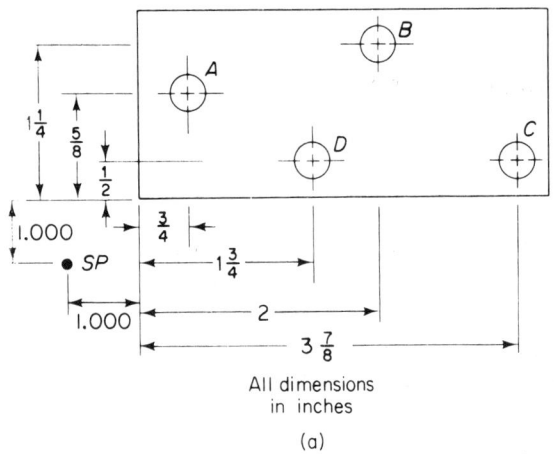

(a)

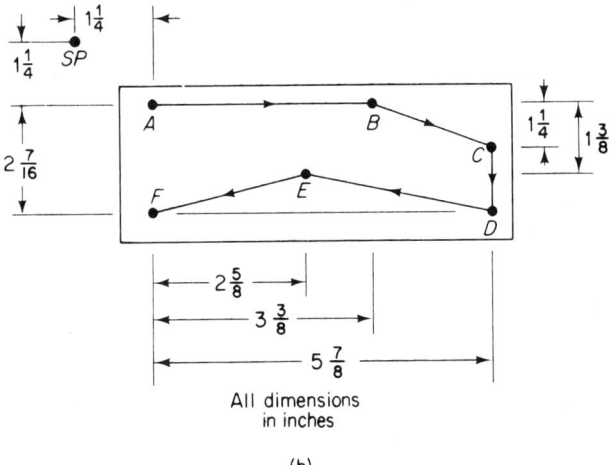

(b)

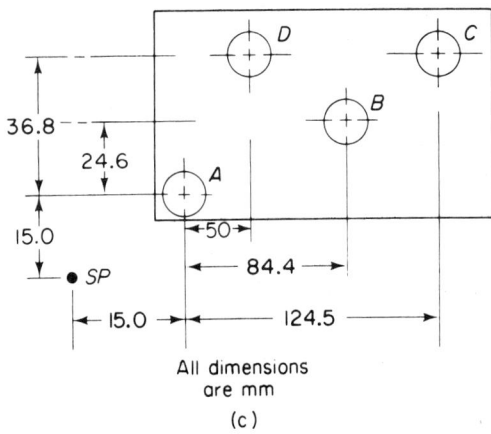

(c)

Figure 7.14

Chap. 7 Questions and Problems

7.18. What is the purpose of the parity channel? Explain.

7.19. (a) How is zero handled in the EIA system?
(b) How is zero handled in the ASCII system?

7.20. (a) How do sprockets read the eight channels on a tape?
(b) How are the holes read in an electronic reader?

7.21. Illustrate the following by punching holes in a tape using the binary-coded-decimal system discussed in this chapter.

Operation	3	Feed	30 in./min
X	0.7460	Speed	90 rev/min
Y	0.3254	End of block	
Z	-0.0562		

7.22. Write the program and show the tape for the binary-coded-decimal system discussed in this chapter for the following.

Operation	6	Feed	20 in./min
X	-0.654	Speed	60 rev/min
Y	0.105	End of block	
Z	0.475		

7.23. Make a diagram of a milling machine table and indicate the plus and minus directions used.

7.24. Write the sequence of moves needed to drill the four holes in Fig. 7.14(a).

7.25. Write the sequence of moves needed to trace the path shown in Fig. 7.14(b).

7.26. Write the sequence of movement in Fig. 7.14(c).

7.27. Write the sequence of movement in Fig. 7.15.

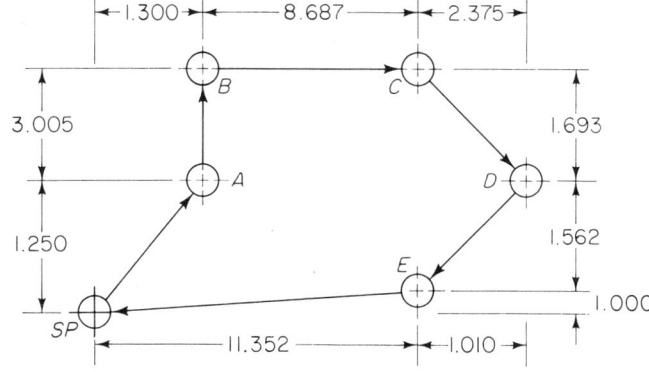

Figure 7.15

112 Incremental CNC   Chap. 7

7.28. Write the program for Fig. 7.16.

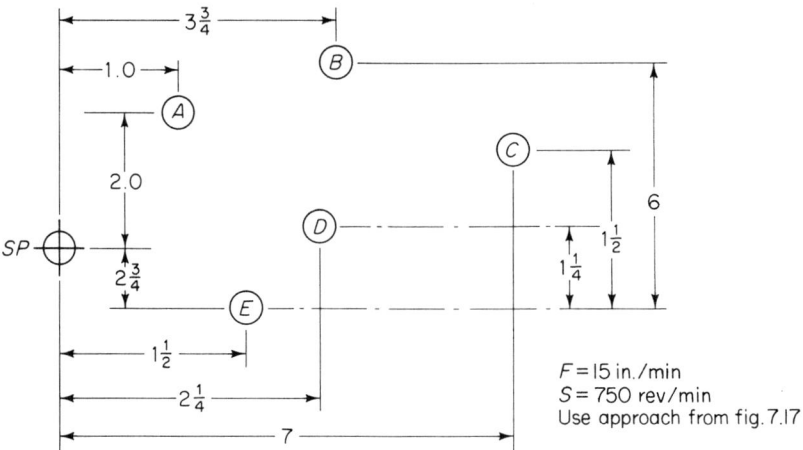

Figure 7.16

7.29. Write the program for Fig. 7.17.

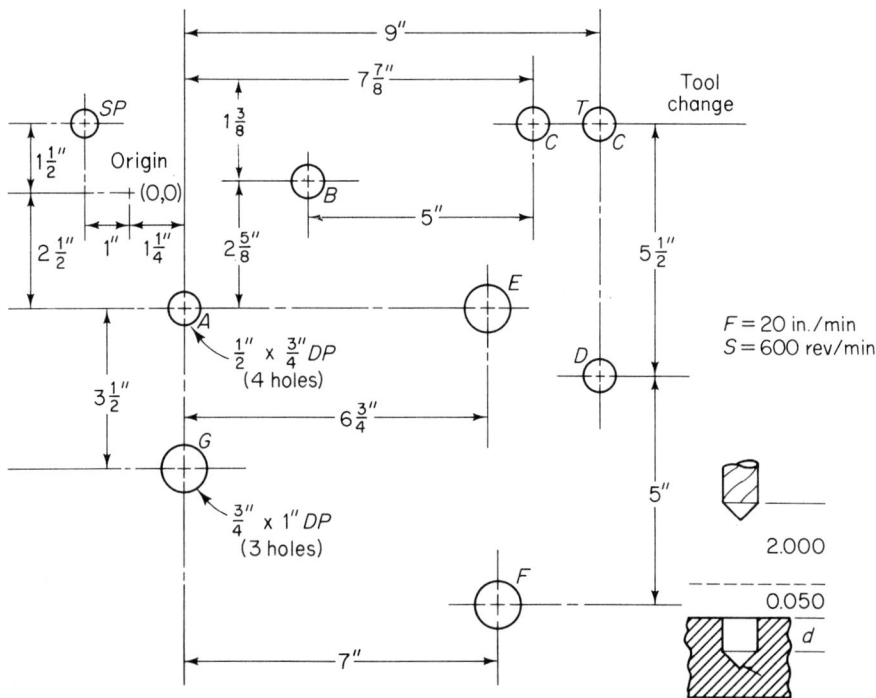

Figure 7.17

Chap. 7 Questions and Problems 113

7.30. Write the program for Fig. 7.18.

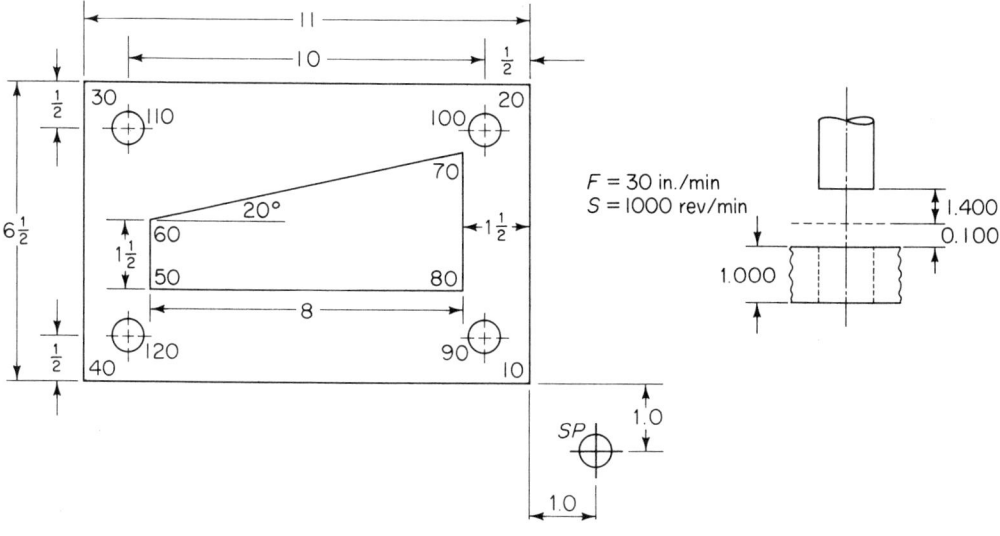

Figure 7.18

7.31. Write the program for Fig. 7.19.

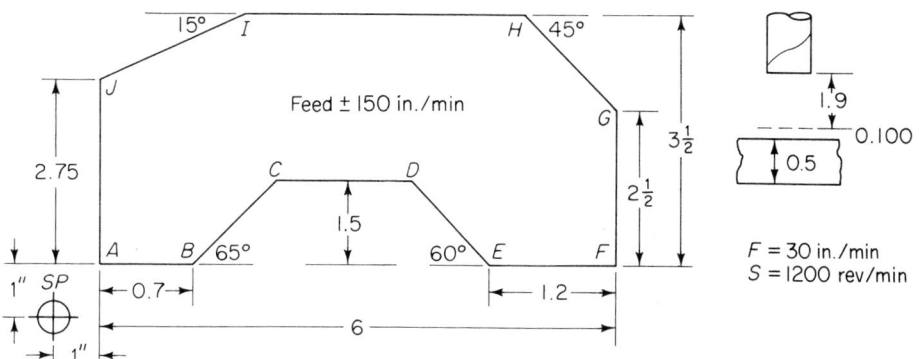

Figure 7.19

7.32. Write the program for Fig. 7.20.

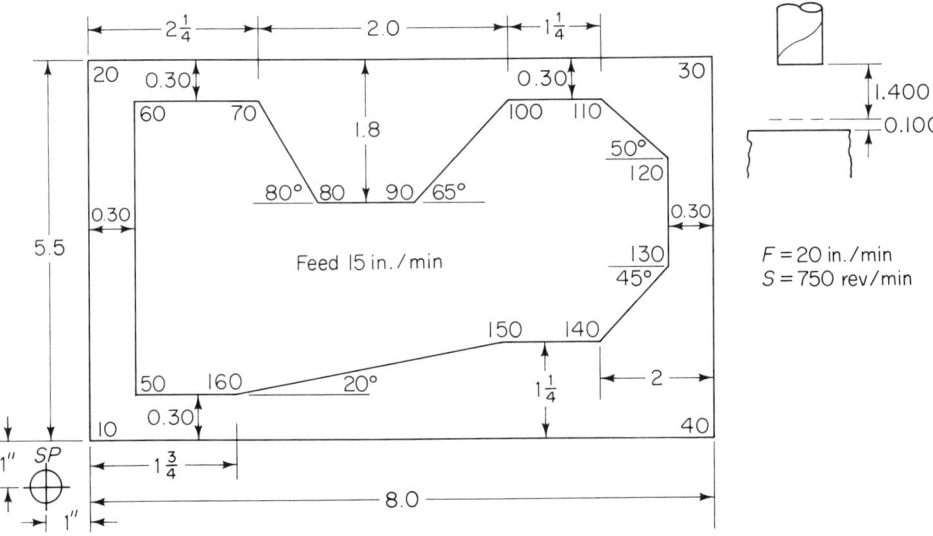

Figure 7.20

7.33. Write the program for Fig. 7.21. The feed is 15 in./min. The speed is 600 rev/min.

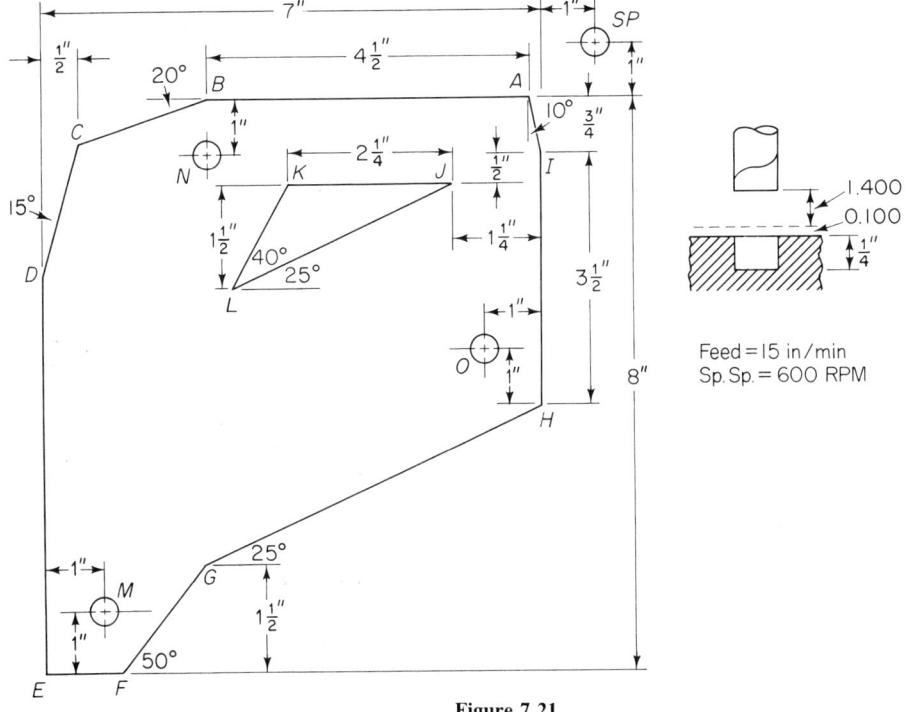

Figure 7.21

Chap. 7   Questions and Problems

7.34. Make a sketch of Fig. 7.22.
   (a) Calculate the coordinates at each point.
   (b) Insert them into your new sketch.
   (c) Program the part. The feed is 30 in./min and the holes are to be milled with a two-lip end mill to a depth of $\frac{1}{4}$ in. The spindle speed is 240 rev/min.

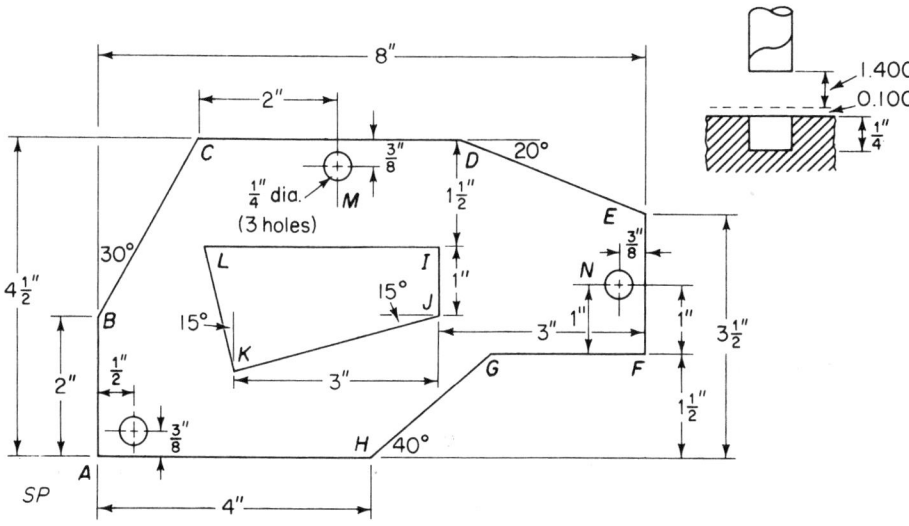

Figure 7.22

8
Circular Contouring: Incremental Mode

8.1 CIRCULAR CONTOURING

Circular contouring is a process used to move a cutter in a circular path. In the process presented in this chapter, each of the four quadrants is programmed separately. In a subsequent chapter, other programming processes will be introduced.

In this system the X and Y linear movements and the I and J movements are programmed on the same line. If only an X and Y movement is programmed, a straight line results. The I and J commands cause an arc to be generated. In this process the *start point of the arc* is defined relative to the *center of the arc*.

Thus, in Fig. 8.1, the X direction to the right is plus (+); the Y direction down is minus (−). The I and J directions are both plus (+). In this system it is important to note that three codes have the same sign when I and J are programmed and the fourth code has the opposite sign. In Fig. 8.1 three codes are plus (+) and one is minus (−).

Consider Fig. 8.2. The movement is clockwise from *A* to *B*. In the X direction the movement is +1.500 and in the Y direction it is −1.500. The I and J movements are taken *from the center of rotation along the X and Y axes to the start point of the arc*. Therefore, from the center of rotation ⊕ to the start point of the radius at *A*, the distance along the X axis is zero (0.000). Along the Y axis the distance is +1.500. The zero X distance is plus (+). The program line from

Sec. 8.1 Circular Contouring

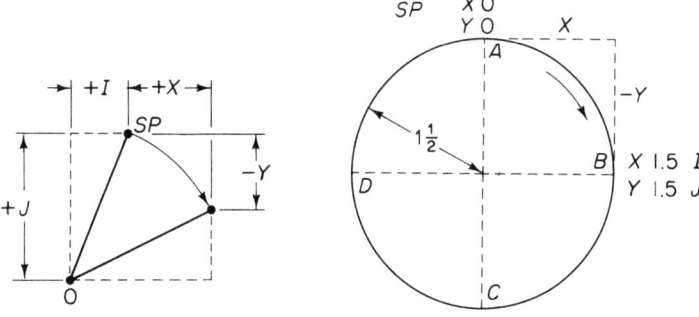

Figure 8.1

Figure 8.2

A to *B* is

X+1.500   Y-1.500   I+0.000   J+1.500

It should be noted that in some systems five digits are programmed. The decimal point is presumed to be between the third and fourth digits. Thus 1.500 is programmed 01500. In this book we will retain the zeros for purposes of analysis [see Table 8.2(a)]. The zeros will be dropped in the programs [see Table 8.2(b)].

The student should verify the movements from *A* to *B*, *B* to *C*, *C* to *D*, and *D* to *A* in Fig. 8.2, as shown in Table 8.1.

Example 1

(a) Do the analysis for Fig. 8.3.

(b) Use a feed of 25 in./min, a spindle speed of 250 rev/min, and a depth of 0.250 in. into the work. Write the program.

Solution (a) The analysis is shown in Table 8.2(a).

(b) The program is shown in Table 8.2(b). In the program, at N30, the tool rapid feeds (G00) to within 0.100 in. of the surface of the work. At N40 it feeds into the work at a rate of 25 in./min (G01) for

TABLE 8.1

Move	X	Y	I	J
A to B	+1.500	-1.500	0.000	+1.500
B to C	-1.500	-1.500	+1.500	0.000
C to D	-1.500	+1.500	0.000	-1.500
D to E	+1.500	+1.500	-1.500	0.000

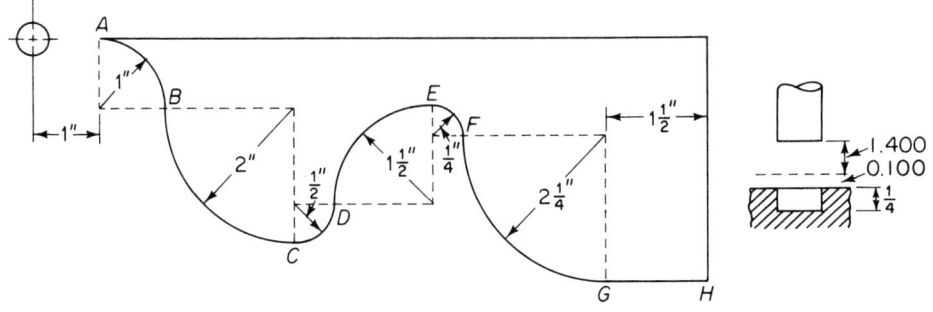

Figure 8.3

a distance of 0.350 in. (0.100 + 0.250). At N120 the quill retracts at a rapid traverse (G00) for a distance of 1.75 in. (1.400 + 0.350).

Example 2

Analyze Fig. 8.4(a) in the incremental mode.

Solution In Fig. 8.4(b),

$$\sin 20° = \frac{x_1}{2}$$

$$x_1 = 2 \sin 20° = 0.684$$

$$\cos 20° = \frac{y_1}{2}$$

$$y_1 = 2 \cos 20° = 1.879$$

In Fig. 8.4(c),

$$\sin 30° = \frac{y_2}{2}$$

$$y_2 = 2 \sin 30° = 1.000$$

$$\cos 30° = \frac{x_2}{2}$$

$$x_2 = 2 \cos 30° = 1.732$$

Therefore,

$$x = x_2 - x_1 = 1.732 - 0.684 = 1.048$$

$$y = y_1 - y_2 = 1.879 - 1.000 = 0.879$$

The analysis from A to B is [Fig. 8.4(d)]

    X1.048   Y-0.879   I0.684   J1.879

TABLE 8.2(a)

Move	X	Y	I	J
SP to A	1.000			
A to B	1.000	-1.000	0.000	1.000
B to C	2.000	-2.000	-2.000	0.000
C to D	0.500	0.500	0.000	-0.500
D to E	1.500	1.500	-1.500	0.000
E to F	0.250	-0.250	0.000	0.250
F to G	2.250	-2.250	-2.250	0.000
G to H	1.500			
H to SP	-10.000	3.500		

TABLE 8.2(b)

N Address	G Code[a]	X Axis	Y Axis	Z Axis	I Arc	J Arc	Spindle Speed (rev/min)	Feed (in./min)	M Function	Comment
N10							250		M03	Spindle on: CW
N20	G00	1.0								SP to A
N30				-1.4						At A
N40	G01			-0.35				25.0		0.025 into work
N50	G02	1.0	-1.0		0.0	1.0				A to B
N60	G03	2.0	-2.0		-2.0	0.0				B to C
N70	G03	0.5	0.5		0.0	-0.5				C to D
N80	G02	1.5	1.5		-1.5	0.0				D to E
N90	G02	0.25	-0.25		0.0	0.25				E to F
N100	G03	2.25	-2.25		-2.25	-0.0				F to G
N110	G01	1.5								G to H
N120	G00			1.75						At H
N130		-10.0	3.5							H to SP
N140									M02	(end of program)

[a] G00, rapid traverse; G01, linear feed; G02, clockwise arc; G03, counterclockwise arc.

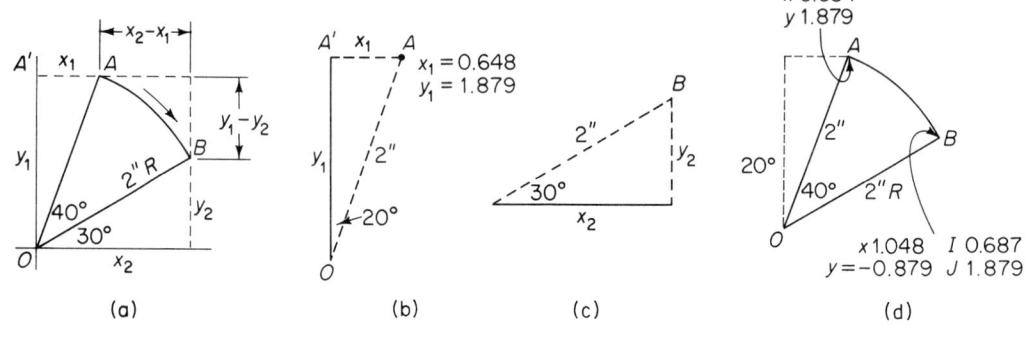

Figure 8.4

Example 3

(a) Analyze the movements for Fig. 8.5(a).
(b) Program Fig. 8.5(a) using a feed of 15 in./min and a spindle speed of 400 rev/min.

Solution (a) The analysis for Fig. 8.5(a) follows: Curve A to E represents the centerline of an end mill.

A to B: Figure 8.5(b) is the arc A to B. A perpendicular is constructed from AO to B, generating triangle $OA'B$. The hypotenuse is 1 in. The movement from A to B is [Fig. 8.5(b)]

$$\sin 45° = \frac{A'B}{1.000}$$

$$A'B = 1.000 \sin 45° = 0.707 \text{ in.}$$

$$\cos 45° = \frac{OA'}{1.000}$$

$$OA' = 1.000 \cos 45° = 0.707 \text{ in.}$$

The Y movement from A to A' is

$$AA' = OA - OA' = 1.000 - 0.707 = 0.293 \text{ in.}$$

The blocks for the A to B move are

X0.707   Y0.293   I0.000   J-1.000

B to C: The movement from B to C [Fig. 8.5(c)] requires the perpendicular BC'. From the triangle $OC'B$, the calculations are similar to those for $OA'B$ [Fig. 8.5(b)]. Therefore, the blocks for the move BC are shown in Fig. 8.5(c). They are

X0.700   Y0.293   I-0.707   J0.707

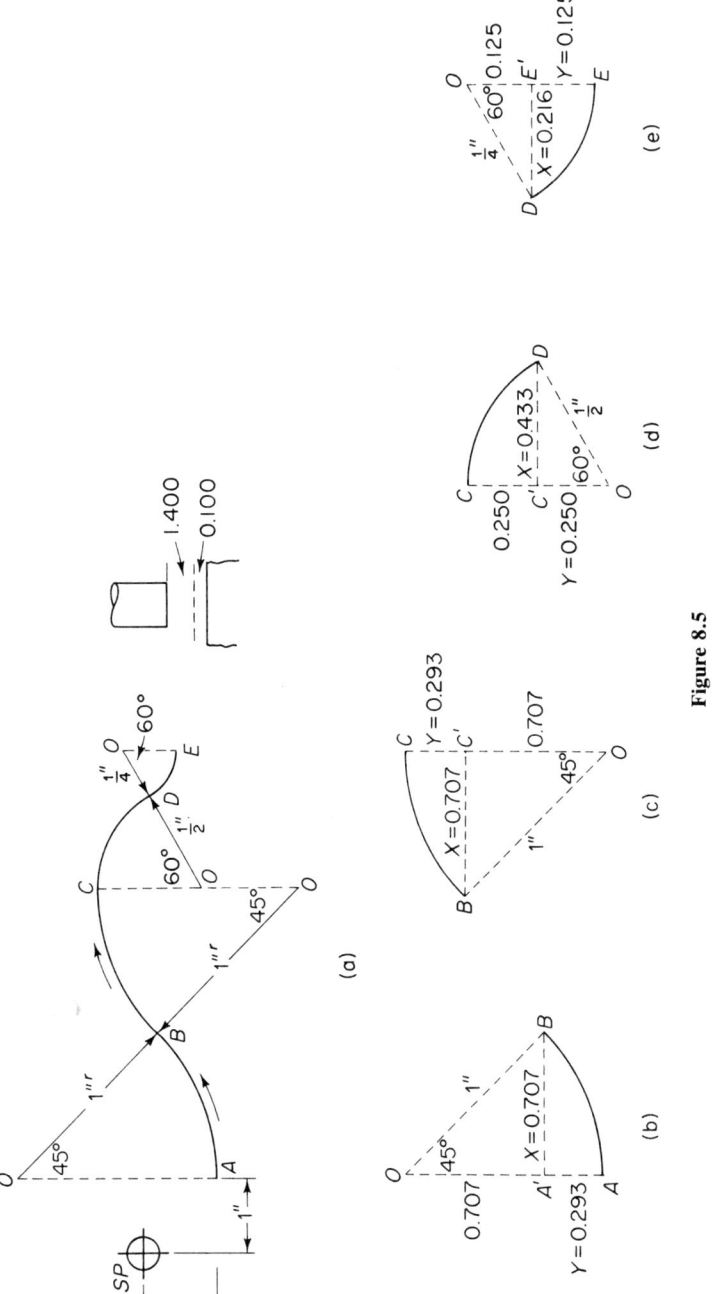

Figure 8.5

C to *D*: The movement from *C* to *D* requires the perpendicular *DC'* in triangle *OC'D* [Fig. 8.5(d)]. From triangle *OC'D* [Fig. 8.5(d)]

$$\sin 60° = \frac{C'D}{0.500}$$

$$X = C'D = 0.500 \sin 60° = 0.433 \text{ in.}$$

$$\cos 60° = \frac{OC'}{0.500}$$

$$OC' = 0.500 \cos 60° = 0.250 \text{ in.}$$

Therefore,

$$Y = CC' = 0.500 - 0.250 = 0.250 \text{ in.}$$

The blocks for the *C*-to-*D* movement in Fig. 8.5(d) are

X0.433   Y-0.250   I0.000   J0.500

D to *E*: The movement from *D* to *E* in Fig. 8.5(e) from triangle *ODE'* is

$$\sin 60° = \frac{DE'}{0.250}$$

$$X = DE' = 0.250 \sin 60° = 0.216 \text{ in.}$$

$$\cos 60° = \frac{OE'}{0.250}$$

$$OE' = 0.250 \cos 60° = 0.125 \text{ in.}$$

Therefore,

$$Y = 0.250 - 0.125 = 0.125 \text{ in.}$$

The blocks for the *D*-to-*E* movement are

X0.216   Y-0.125   I-0.125   J-0.125

E to SP: The movement is

$$X = -0.216 - 0.433 - 0.707 - 0.707 - 1.000$$
$$= -3.063 \text{ in.}$$
$$Y = 0.125 + 0.250 - 0.293 - 0.293 + 1.000$$
$$= 0.789 \text{ in.}$$

The blocks for the *E*-to-SP move are

X-3.063   Y0.789

TABLE 8.3(a)

Move	X	Y	Z	I	J
SP to A	1.000	-1.000			
At A			-1.400		
At A			-0.100		
A to B	0.707	0.293		0.000	-1.000
B to C	0.707	0.293		-0.707	0.707
C to D	0.433	-0.250		0.000	0.500
D to E	0.216	-0.125		-0.216	-0.125
At E			1.500		
E to SP	-3.063	0.789			

TABLE 8.3(b)

N Address	G Code[a]	X Axis	Y Axis	Z Axis	I Arc	J Arc	Spindle Speed (rev/min)	Feed (ft/min)	M Function	Comment
N10							400		M03	Spindle on: CW
N20	G00	1.0	-1.0							SP to A
N30				-1.4						At A
N40	G01			-0.1						At A
N50	G03	0.707	0.293		0.0	-1.0		15.0		A to B
N60	G02	0.707	0.293		-0.707	0.707				B to C
N70	G02	0.433	-0.250		0.0	0.5				C to D
N80	G03	0.216	-0.125		-0.216	-0.125				D to E
N90	G00			1.5						At E
N100		-3.063	0.789							E to SP
N110									M02	(end of program)

[a]G00, rapid traverse; G01, linear feed; G02, clockwise arc; G03, counterclockwise arc.

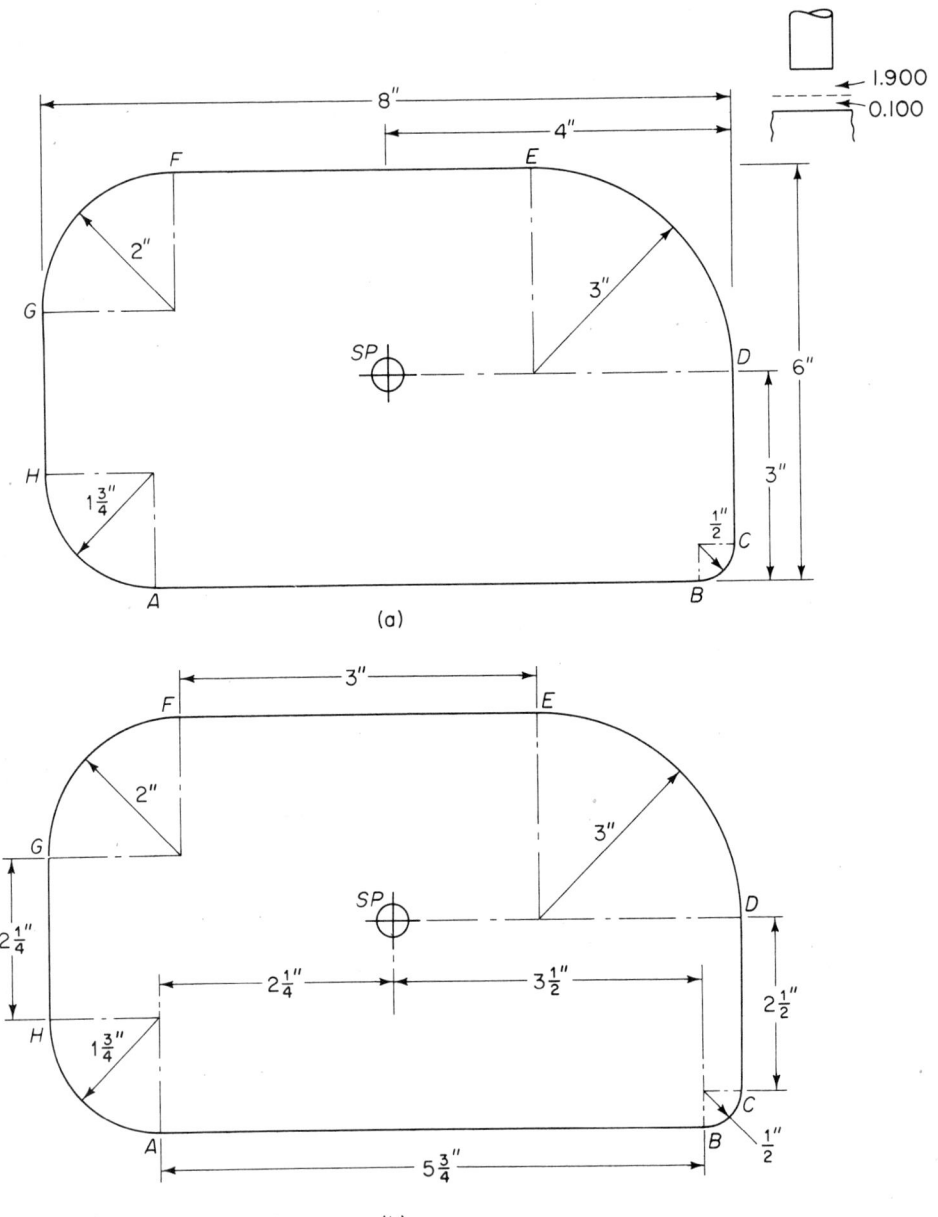

Figure 8.6

Sec. 8.1 Circular Contouring

The summary analysis for Fig. 8.5(a) is shown in Table 8.3(a). The sum of the X and Y columns is zero.

(b) The program for Fig. 8.5 is shown in Table 8.3(b).

Example 4

(a) Analyze the movements in Fig. 8.6(a).

(b) Write the program using a feed of 15 in./min and a spindle speed of 300 rev/min.

Solution From Fig. 8.6(b):

SP to A:

X = −2.250
Y = −3.000

A to B:

X = 5.750
Y = 0.000

B to C:

X = 0.500   I = 0.000
Y = 0.500   J = −0.500

C to D:

X = 0.000
Y = 2.500

D to E:

X = −3.000   I = 3.000
Y = 3.000   J = 0.000

E to F:

X = −3.000
Y = 0.000

F to G:

X = −2.000   I = −0.000
Y = −2.000   J = 2.000

G to H:

X = 0.000
Y = −2.250

H to A:

X = 1.750   I = −1.750
Y = −1.750   J = −0.000

A to SP:

X = 2.250
Y = 3.000

The composite analysis for Fig. 8.6(b) is shown in Table 8.4(a).

The algebraic sum of the X column should equal zero, and the algebraic sum of the Y column should equal zero.

(b) The program for Fig. 8.6(a) is shown in Table 8.4(b).

TABLE 8.4(a)

Move	X	Y	I	J	Z
SP to A	-2.250	-3.000			
At A					-1.900
At A					-0.100
A to B	5.750	0.500	0.000	-0.500	
B to C	0.500	2.500			
C to D	0.000	3.000	3.000	0.000	
D to E	-3.000	0.000			
E to F	-3.000	-2.000	0.000	2.000	
F to G	-2.000	-2.250			
G to H	0.000	-1.750	-1.750	0.000	
H to A	1.750				
At A					2.000
A to SP	2.250	3.000			

TABLE 8.4(b)

N Address	G Code[a]	X Axis	Y Axis	Z Axis	I Arc	J Arc	Spindle Speed (rev/min)	Feed (in./min)	M Function	Comment
N10							300		M03	Spindle on: CW
N20	G00	-2.75	-3.0							SP to A
N30				-1.9						At A
N40	G01			-0.1				15.0		At A
N50		5.75	0.5							A to B
N60	G03	0.5	2.5		0.0	-0.5				B to C
N70	G01									C to D
N80	G03	-3.0	3.0		3.0	0.0				D to E
N90	G01	-3.0	-2.0							E to F
N100	G03	-2.0	-2.25		0.0	2.0				F to G
N110	G01		-1.75							G to H
N120	G03	1.75			-1.75	0.0				H to A
N130	G00			2.0						At A
N140		2.25	3.0							A to SP
N150									M02	(end of program)

[a]G00, rapid traverse; G01, linear feed; G03, counterclockwise.

Chap. 8 Questions and Problems

QUESTIONS AND PROBLEMS

8.1. What effect do the I and J commands have on the movement of the cutter?

8.2. What is the significance of the center of the radius when determining the I and J values?

8.3. Explain:
  (a) The G08 and the G09 codes.
  (b) The M10, the M20, and the M30 codes.

8.4. Write the program for Fig. 8.7(a). All programs are to be written in the incremental mode.

8.5. Write the program for Fig. 8.7(b).

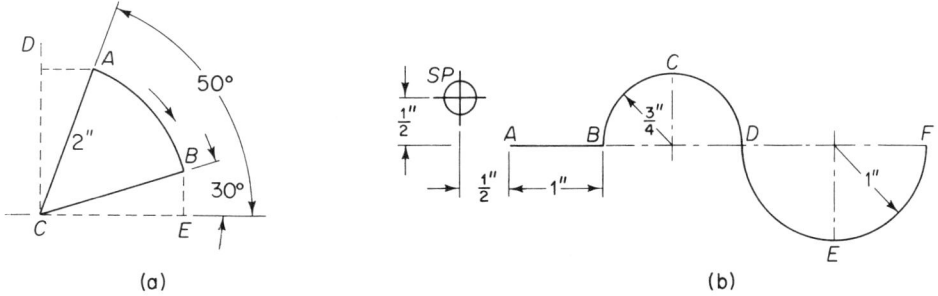

Figure 8.7

8.6. Write the program for Fig. 8.8(a).
8.7. Write the program for Fig. 8.8(b).

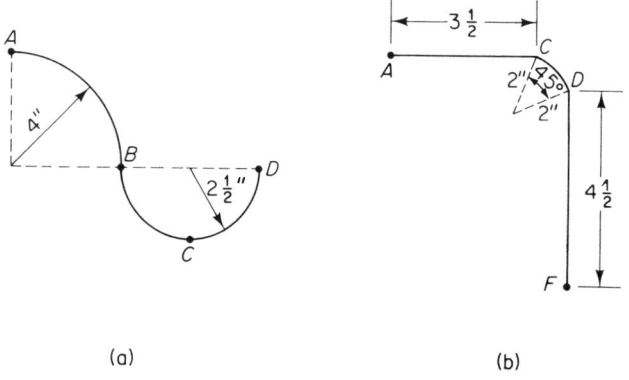

Figure 8.8

8.8. Write the program for Fig. 8.9. The feed is 20 in./min and the spindle speed is 300 rev/min.

128   Circular Contouring: Incremental Mode   Chap. 8

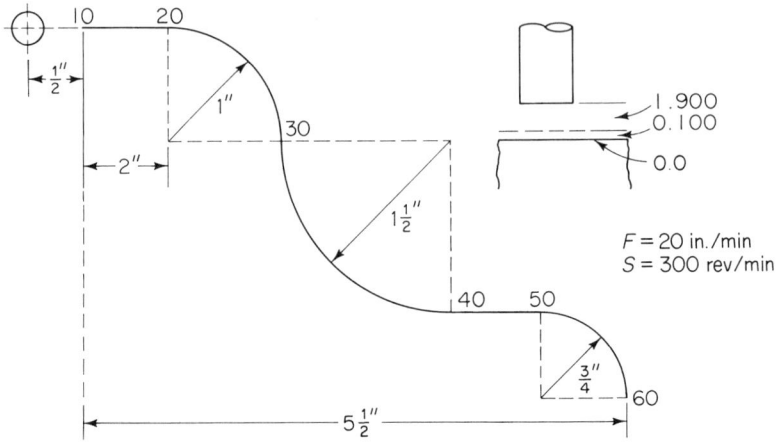

Figure 8.9

8.9. Write the program for Fig. 8.10. The feed is 15 in./min and the spindle speed is 250 rev/min.

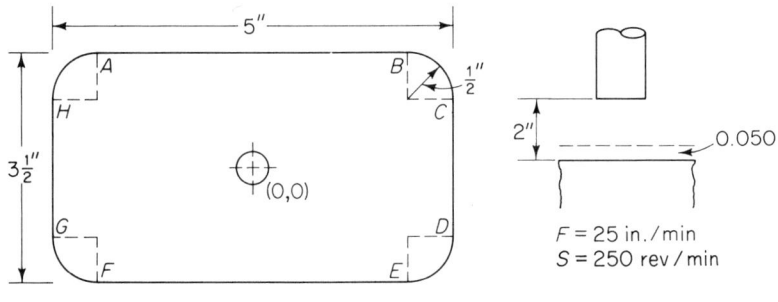

Figure 8.10

8.10. Write the program to trace the outline of the work in Fig. 8.11. The approach is $\frac{3}{8}$ in. The feed is 20 in./min. The spindle speed is 450 rev/min.

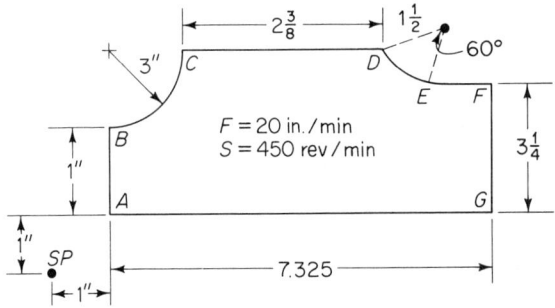

Figure 8.11

Chap. 8 Questions and Problems

8.11. Write the program for Fig. 8.12. The feed is 15 in./min and the spindle speed is 750 rev/min.

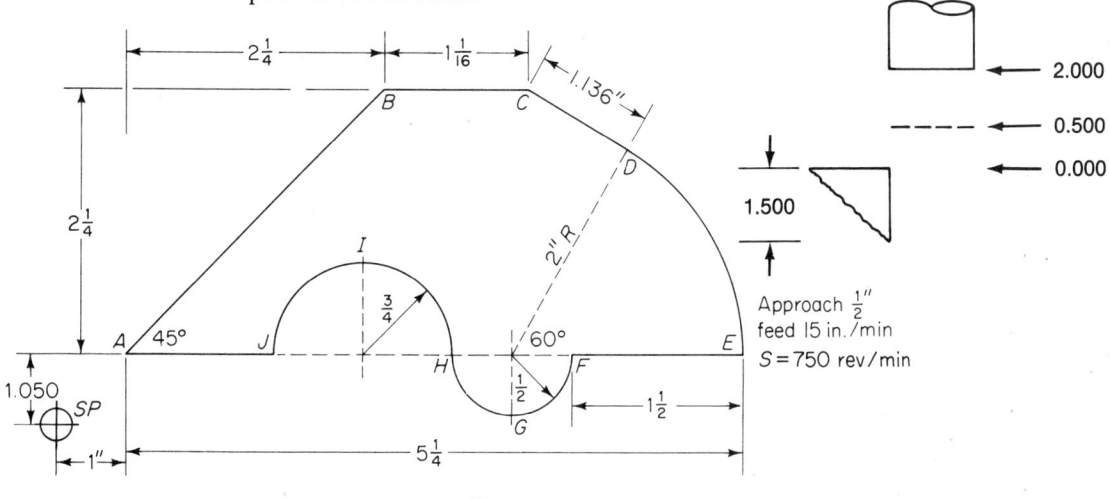

Figure 8.12

8.12. Write the program for Fig. 8.13. The feed is 30 in./min and the spindle speed is 600 rev/min.

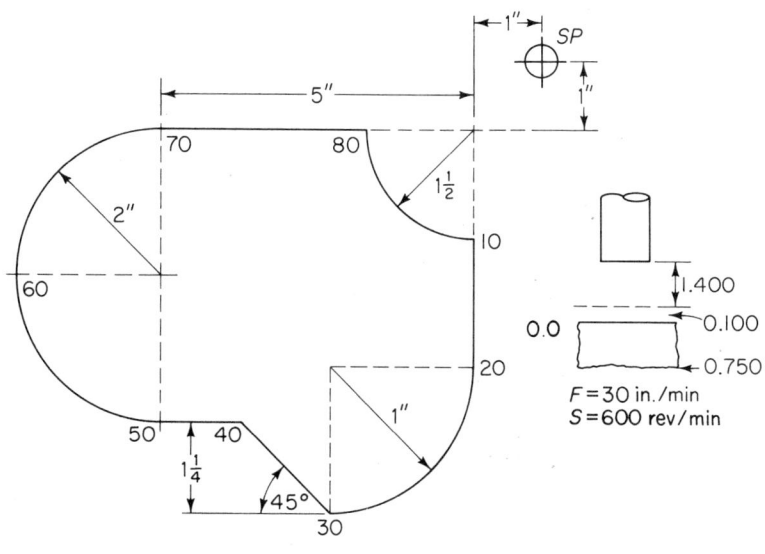

Figure 8.13

8.13. Write the program for Fig. 8.14. The feed is 15 in./min. The spindle speed is 700 rev/min.

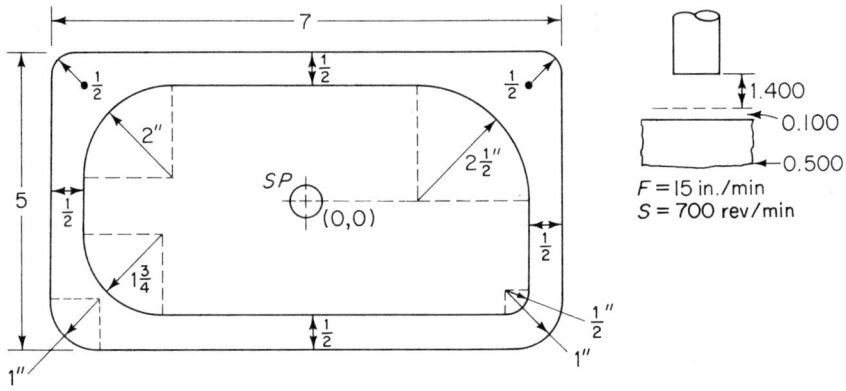

Figure 8.14

8.14. Write the program for Fig. 8.15. The feed is 15 in./min. The spindle speed is 300 rev/min.

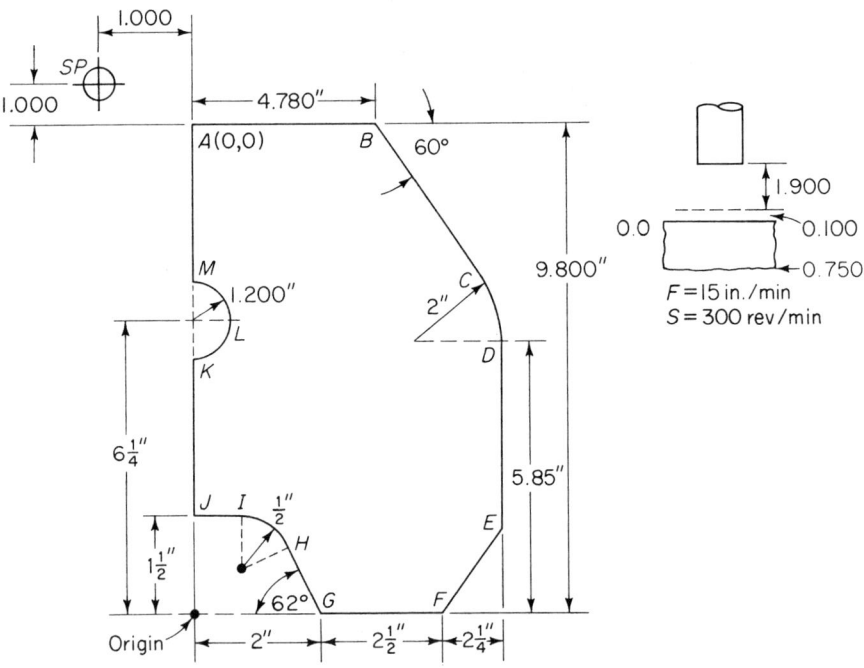

Figure 8.15

Chap. 8   Questions and Problems

8.15. Write the program for Fig. 8.16. The feed is 20 in./min. The spindle speed is 425 rev/min.

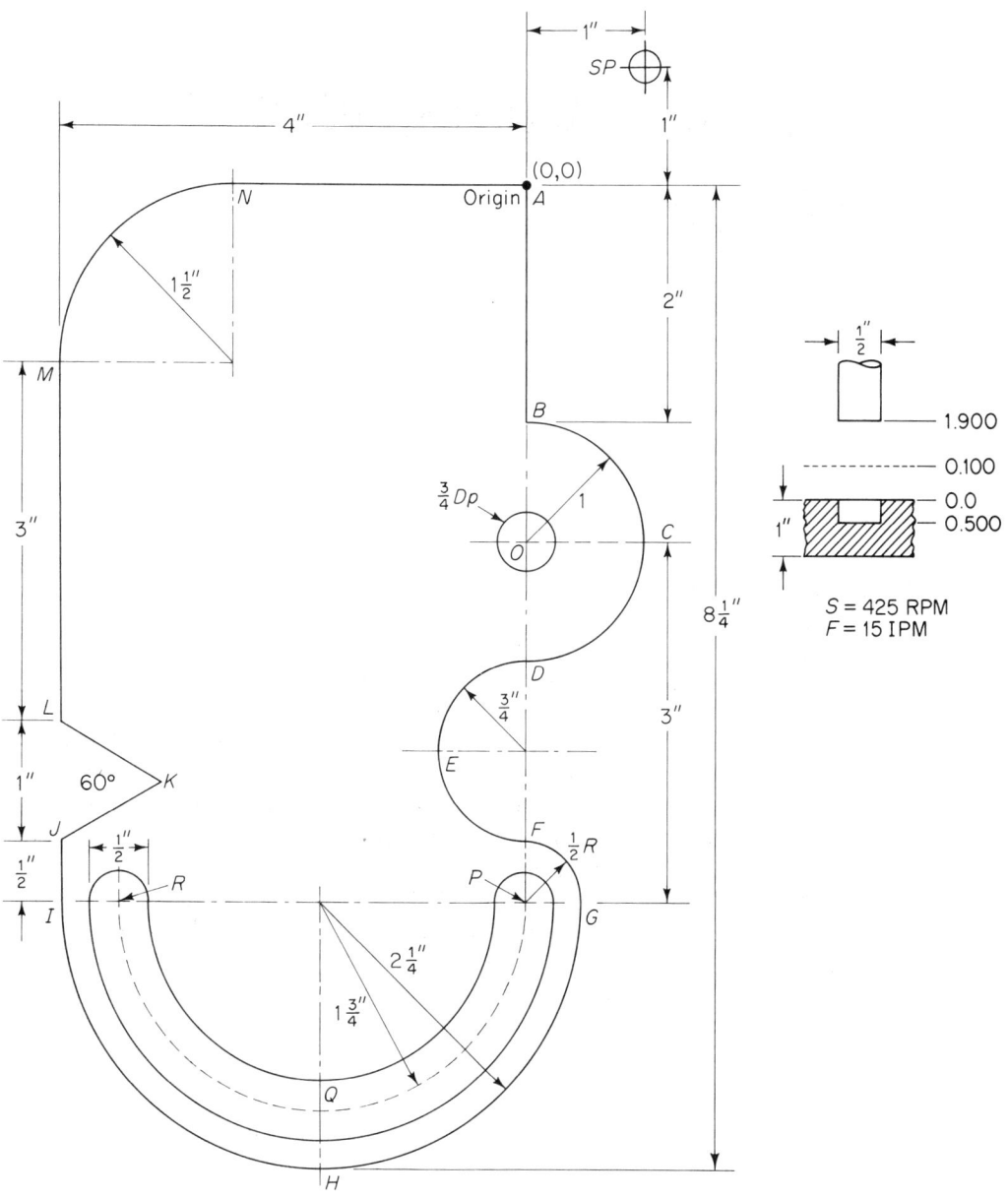

Figure 8.16

9

Absolute Mode: Milling Machine

9.1 ABSOLUTE SYSTEM

In the preceding chapters, the movement from one point to the next point was independent from all other points. Each starting point was taken as zero. This was characterized as an *incremental system*. We are now going to consider a system where the movement of the cutter is an algebraic summation of all points in the X direction, the Y direction, or the Z dimension. Therefore, as the plus and minus distances are added, the return to the origin should sum to zero in the three coordinate directions. This is called the *absolute system*. The absolute movements for Fig. 9.1(a) are shown in Fig. 9.1(b). The absolute movements in Fig. 9.1(b) should be compared with the incremental movements in Fig. 7.10(b).

A to *B*:
$$X = -1.500$$
$$Y = -3.000$$

B to *C*:
$$X = -1.500 - 2.250 = -3.750$$
$$Y = -3.000 + 1.750 = -1.250$$

C to *D*:
$$X = -3.750 - 4.375 = -8.125$$
$$Y = -1.250 - 1.250 = -2.500$$

Sec. 9.1 Absolute System

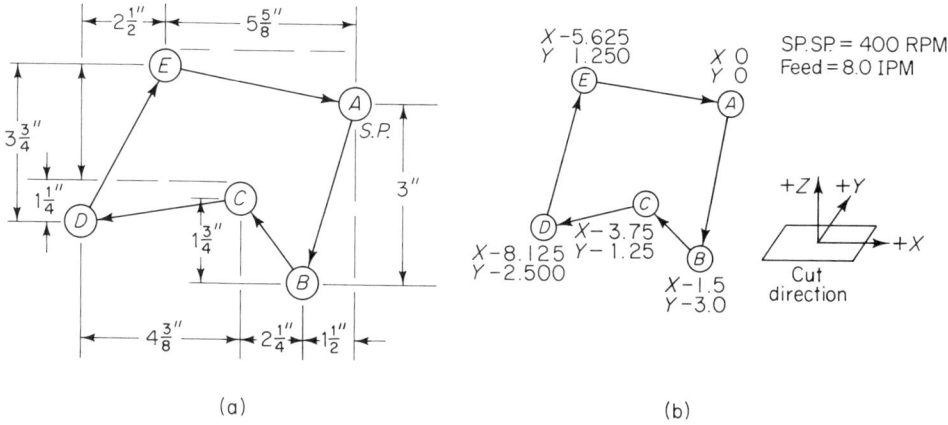

Figure 9.1

D to *E*:

$$X = -8.125 + 5.625 = -5.625$$
$$Y = -2.500 + 3.750 = 1.250$$

E to *A*:

$$X = -5.625 + 5.625 = 0.000$$
$$Y = -1.250 - 1.250 = 0.000$$

Again, each new move adds algebraically to the previous sum. The end result is X = 0 and Y = 0. The analysis is as shown in Table 9.1(a). Note that leading and trailing zeros need not be programmed. This is shown in Fig. 9.1(a).

The student is urged to redraw the configuration and insert the values to be programmed. This is shown in Fig. 9.1(b).

The program for Fig. 9.1(b) is shown in Table 9.1(b).

Example 1

(a) Determine the X and Y values at each junction point in Fig. 9.2(a) and insert these values in Fig. 9.2(b).
(b) Do the analysis for Fig. 9.2(b).
(c) Write the program in the absolute mode. Use a feed of 80 in./min and a spindle speed of 400 rev/min.

Solution (a) The start point is at SP. The origin is at *A*. The cutter retreats to SP at the end of each machine cycle. All dimensions and

TABLE 9.1(a)

N	X	Y	Move
N10	-1.500	-3.000	A to B
N20	-3.750	-1.250	B to C
N30	-8.125	-2.500	C to D
N40	-5.625	1.250	D to E
N50	0.000	0.000	E to A

TABLE 9.1(b)

N Address	G Code	X Axis	Y Axis	Spindle Speed (rev/min)	Feed (ft/min)	M Function	Comment
N10				400		M03	Turn spindle on: CW
N20	G01	-1.5	-3.0		8.0		Linear feed, A to B
N30		-3.75	-1.25				B to C
N40		-8.125	-2.5				C to D
N50		-5.625	1.25				D to E
N60		0.0	0.0				E to A
N70						M02	(end of program)

Sec. 9.1   Absolute System   135

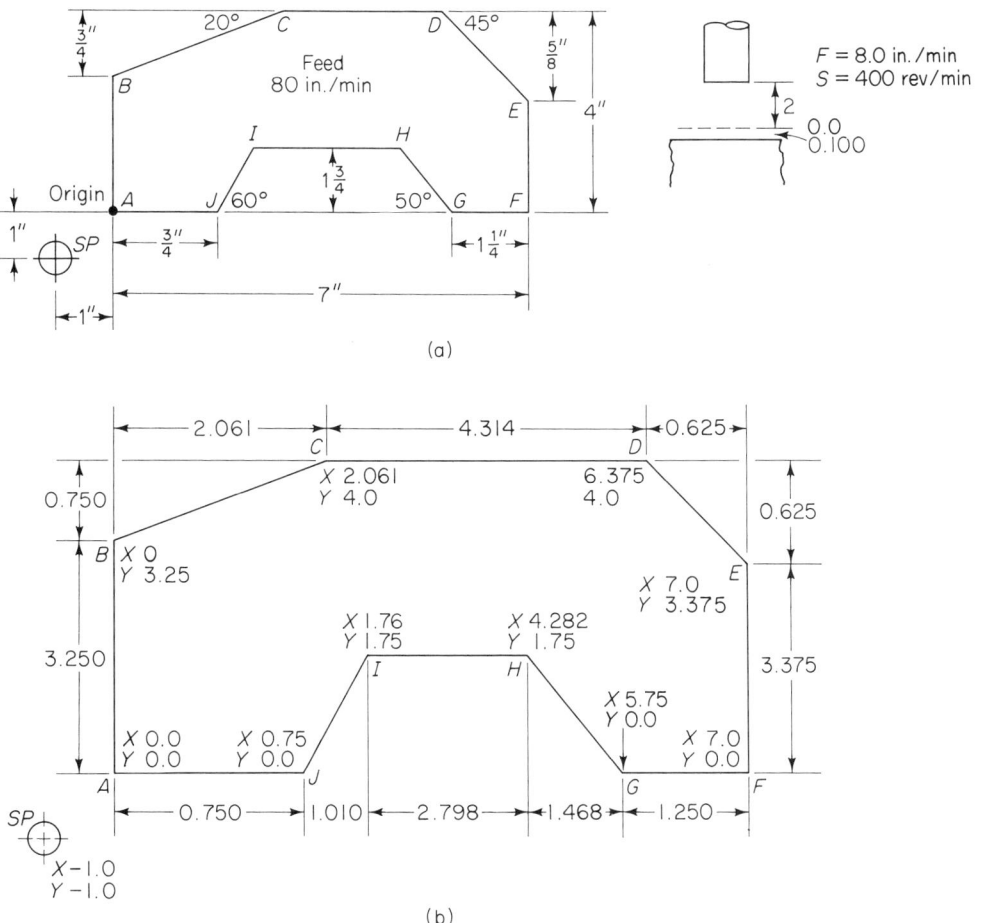

Figure 9.2

movements are calculated from the origin. Therefore, at A,

$$X = 0.000 \text{ in.}$$
$$Y = 0.000 \text{ in.}$$

Moving in a clockwise direction, the next directional change is at B.
A to B:

$$X = 0.000 \text{ in.}$$
$$Y = 0.000 + (4.000 - 0.750) = 3.250 \text{ in.}$$

B to *C*:

$$\tan 20° = \frac{0.750}{x}$$

$$x = \frac{0.750}{\tan 20°}$$

$$= 2.061 \text{ in.}$$

$$y = 0.750 \text{ in.}$$

Therefore,

$$X = 0.000 + 2.061 = 2.061 \text{ in.}$$

$$Y = 3.250 + 0.750 = 4.000 \text{ in.}$$

C to *D*:

$$X = 2.061 + (7.000 - 2.061 - 0.625)$$

$$= 6.375 \text{ in.}$$

$$Y = 4.000 \text{ in.}$$

D to *E*:

$$X = 6.375 + 0.625 = 7.000 \text{ in.}$$

$$Y = 4.000 - 0.625 = 3.375 \text{ in.}$$

E to *F*:

$$X = 7.000 \text{ in.}$$

$$Y = 3.375 - 3.375 = 0.000 \text{ in.}$$

F to *G*:

$$X = 7.000 - 1.250 = 5.750 \text{ in.}$$

$$Y = 0.000 \text{ in.}$$

G to *H*:

$$\tan 50° = \frac{1.750}{x}$$

$$x = \frac{1.750}{\tan 50°}$$

$$= 1.468 \text{ in.}$$

Sec. 9.1   Absolute System   137

Therefore,

$$X = 5.750 - 1.468 = 4.282 \text{ in.}$$
$$Y = 0.000 + 1.750 = 1.750 \text{ in.}$$

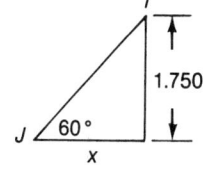

H to *I*:

$$\tan 60° = \frac{1.750}{x}$$

$$x = \frac{1.750}{\tan 60°}$$

$$= 1.010 \text{ in.}$$

Therefore,

$$X = 4.282 - (7.000 - 0.750 - 1.010 - 1.468 - 1.250)$$
$$= 4.282 - 2.522 = 1.760 \text{ in.}$$
$$Y = 1.750 \text{ in.}$$

I to *J*:

$$X = 1.760 - 1.010 = 0.750 \text{ in.}$$
$$Y = 1.750 - 1.750 = 0.000$$

J to *A*:

$$X = 0.750 - 0.750 = 0.000 \text{ in.}$$
$$Y = 0.000$$

A to SP:

$$X = 0.000 - 1.000 = -1.000 \text{ in.}$$
$$Y = 0.000 - 1.000 = -1.000 \text{ in.}$$

(b) The analysis is tabulated in Table 9.2(a).
(c) The program is shown in Table 9.2(b).

The G00 command is a rapid code. The G90 indicates that the program is in the absolute mode. The T1 tells the computer that the tool in the magazine is the tool to be used. The M6 indicates that tool number 1 is requested. M2 ends the program, turns the spindle off, and resets the counter.

In the command, line N30, the X, Y, and Z movements may be programmed on one line. The X and Y movements will occur before

TABLE 9.2(a)

Move	X	Y	Z
SP	-1.000	-1.000	0.000
SP to A	0.000	0.000	-0.100
At A			
A to B	2.061	3.250	
B to C	6.375	4.000	
C to D	7.000	3.375	
D to E		0.000	
E to F	5.750		
F to G	4.282	1.750	
G to H	1.760		
H to I	0.750	0.000	
I to J	0.000		
J to A	-1.000	-1.000	2.000
A to SP			

TABLE 9.2(b)

N Address	G Code	X Axis	Y Axis	Z Axis	Spindle Speed (rev/min)	Feed (in./min)	M Function	Comment
N10							T01 M06	Spindle on: CW
N20	G90 G00	0.0	0.0	0.0	400		M03	SP to A
N30	G01			-0.1		8.0		At A
N40								A to B
N50		2.061	3.25					B to C
N60		6.375	4.0					C to D
N70		7.0	3.375					D to E
N80			0.0					E to F
N90		5.75						F to G
N100		4.282	1.75					G to H
N110		1.76						H to I
N120		0.75	0.0					I to J
N130		0.0						J to A
N140	G00	-1.0	-1.0	0.2				Rapid up at A to SP
N150							M02	(end of program)

138

Sec. 9.1 Absolute System

the Z movement. In the command, line N150, the Z movement takes place before the X and Y movements.

Example 2 uses an end mill to machine the four slots in Fig. 9.3(a). The end mill should rapid to 0.100 in. above the surface and then feed into the work to a depth of $\frac{1}{4}$ in. Once the full depth is achieved, the end mill starts to trace a particular path. It then rapidly retracts to its "home" position.

Example 2

(a) Analyze and tabulate the movement of the tool in Fig. 9.3(a) in the absolute mode. Position 0 is the origin. Insert these dimensions into Fig. 9.3(b).

(b) Program and tabulate these dimensions.

Solution (a) The analysis of the movement in Fig. 9.3(a) is as follows:
Origin (D) to SP ($\oplus$):

$$X = -1.500 \text{ in.}$$
$$Y = 1.250 \text{ in.}$$

SP to A:

$$X = -1.500 + 8.000 = 6.500 \text{ in.}$$
$$Y = 1.250 - 1.250 = 0.000 \text{ in.}$$

A to B:

$$X = 6.500 - 2.500 = 4.000 \text{ in.}$$
$$Y = 0.000$$

B to C:

$$X = 4.000 - 3.000 = 1.000 \text{ in.}$$
$$Y = 0.000$$

C to D:

$$X = 1.000 - 1.000 = 0.000 \text{ in.}$$
$$Y = 0.000 \text{ in.}$$

D to E:

$$X = 0.000 \text{ in.}$$
$$Y = 0.000 - 2.250 = -2.250 \text{ in.}$$

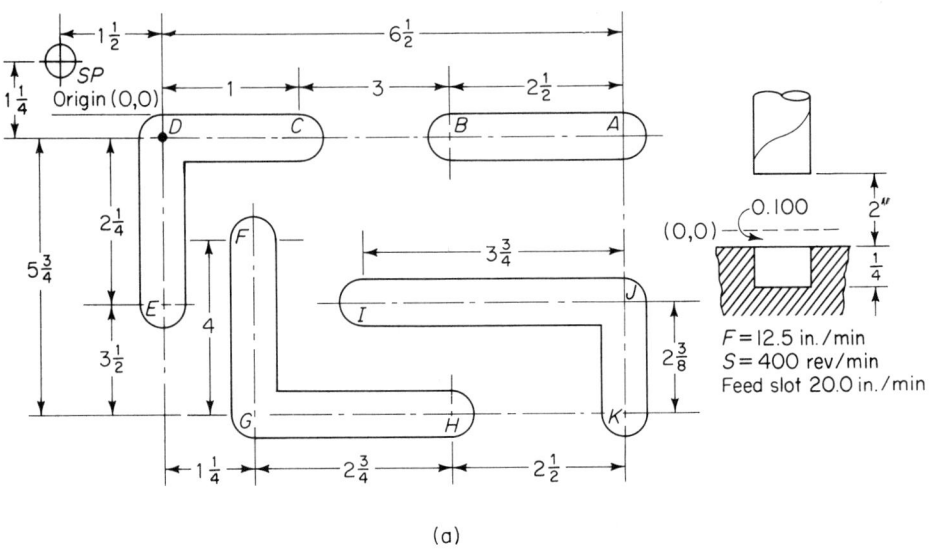

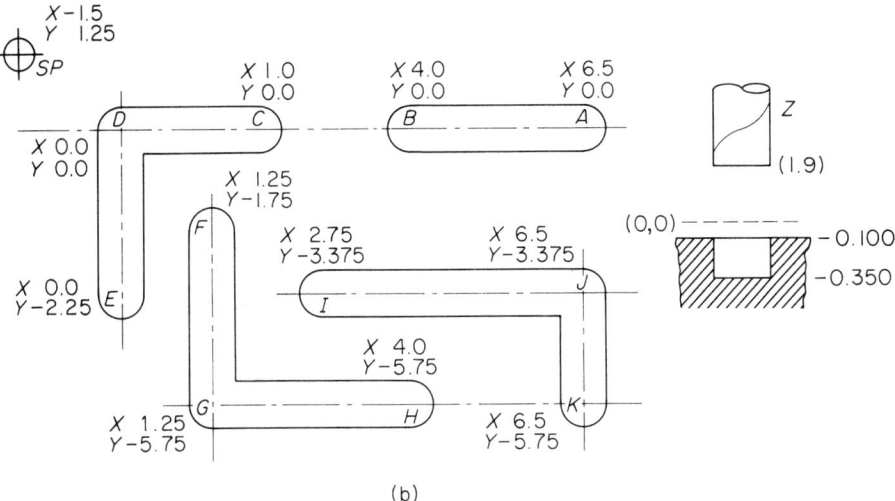

Figure 9.3

Sec. 9.1   Absolute System   141

E to F:

$$X = 0.000 + 1.250 = 1.250 \text{ in.}$$
$$Y = -2.250 + 0.500 = -1.750 \text{ in.}$$

F to G:

$$X = 1.250 \text{ in.}$$
$$Y = -1.750 - 4.000 = -5.750 \text{ in.}$$

G to H:

$$X = 1.250 + 2.750 = 4.000 \text{ in.}$$
$$Y = -5.750 \text{ in.}$$

H to I:

$$X = 4.000 - 1.250 = 2.750 \text{ in.}$$
$$Y = -5.750 + 2.375 = -3.375 \text{ in.}$$

TABLE 9.3(a)

Move	X	Y	Z
SP	-1.500	1.250	
SP to A	6.500	0.000	-1.900
At A			-0.350
A to B	4.000		
At B			0.000
B to C	1.000		
At C			-0.350
C to D	0.000		
D to E		-2.250	
At E			0.000
E to F	1.250	-1.750	
At F			-0.350
F to G		-5.750	
G to H	4.000		
At H			0.000
H to I	2.750	-3.375	
At I			-0.350
I to J	6.500		
J to K		-5.750	
At K			1.900
K to SP	-1.500	1.250	

TABLE 9.3(b)

N Address	G Code	X Axis	Y Axis	Z Axis	Spindle Speed (rev/min)	Feed (ft/min)	M Code	Comment
N10	G00	-1.5	1.25		400		M03	Spindle on: CW SP
N20		6.5	0.0					SP to A
N30								At A
N40				0.0				
N50	G01			-0.35		12.5		Linear feed
N60		4.0				20.0		A to B
N70	G00			0.0				Rapid at B
N80		1.0						B to C
N90	G01			-0.35		12.5		At C
N100		0.0				20.0		C to D
N110			-2.25					D to E
N120	G00			0.0				At E
N130		1.25						E to F
N140	G01		-1.75	-0.35		12.5		At F
N150		4.0				20.0		F to G
N160			-5.75					G to H
N170	G00			0.00				At H
N180		2.75	-3.375					H to I
N190	G01			-0.35		12.5		At I
N200		6.5				20.0		I to J
N210			-5.75					J to K
N220	G00			1.9				At K
N230		-1.5	1.25					K to SP
N240							M02	(end of program)

Chap. 9   Questions and Problems

I to *J*:
$$X = 2.750 + 3.750 = 6.500 \text{ in.}$$
$$Y = -3.375 \text{ in.}$$

J to *K*:
$$X = 6.500 \text{ in.}$$
$$Y = -3.375 - 2.375 = -5.750 \text{ in.}$$

K to SP:
$$X = 6.500 - 8.000 = -1.500 \text{ in.}$$
$$Y = -5.750 + 7.000 = 1.250 \text{ in.}$$

The values above are tabulated in Table 9.3(a) and included in Fig. 9.3(b).

(b) The program is shown in Table 9.3(b).

QUESTIONS AND PROBLEMS

9.1. Describe two methods—incremental and absolute—for programming a machine part.

9.2. Describe the use of:
(a) The two M functions used in Table 9.2(b).
(b) The G codes used in Table 9.2(b).

9.3. Explain the X, Y, Z movements in the program developed for Fig. 9.2(a) in lines N30, N40, and N150.

9.4. Explain the movements in lines N40, N50, N70, and N220 in Table 9.3(b).

9.5. Write the program in the absolute mode for Fig. 9.4. Use a spindle speed of 750 rev/min and a feed of 175 in./min.

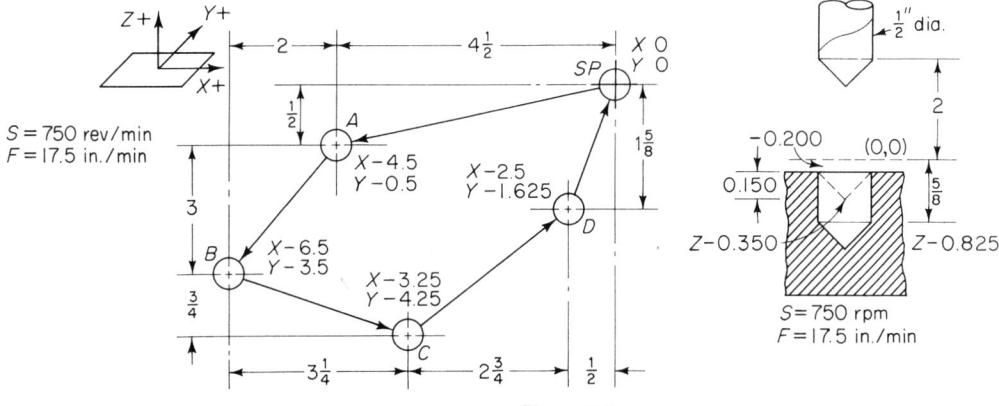

Figure 9.4

9.6. Write the program for Fig. 7.15 in the absolute mode: spindle speed 250 rev/min and a spindle speed of 7.5 in./min at each hole.

9.7. Write the program for Fig. 7.16 in the absolute mode using a feed of 215 in./min between holes and a spindle speed of 20.0 ft/min.

9.8. Repeat Prob. 9.7 for Fig. 7.18. Use a feed of 150 in./min and a spindle speed of 300 rev/min. Use the absolute mode.

9.9. Repeat Prob. 9.7 for Fig. 7.19.

9.10. Repeat Prob. 9.7 for Fig. 7.20.

9.11. Repeat Prob. 9.7 for Fig. 7.21.

9.12. Repeat Prob. 9.7 for Fig. 7.22.

9.13. Program Fig. 9.5 in the absolute mode. Use a 1-in. two-lip end mill. Do not use tool nose radius compensation. Program the center of the cutter.

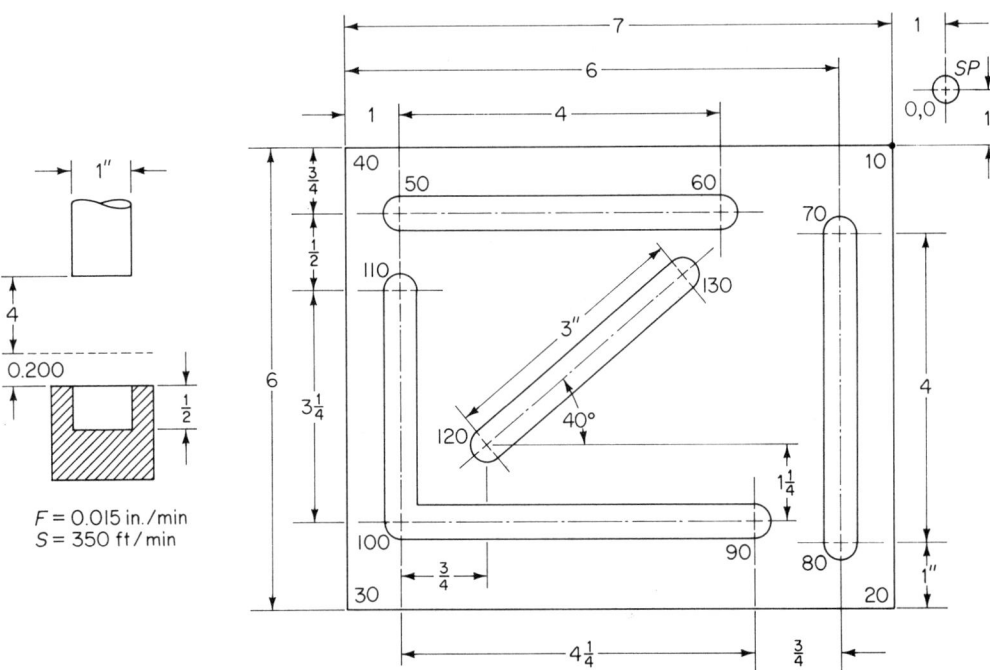

Figure 9.5

10

Linear Interpolation: Drill Routines

10.1 SPOT DRILL

Figure 10.1 requires that four holes be drilled. The procedure requires a spot drill before the holes are drilled. This serves two purposes. It permits the lips of the drill to contact the spot with the lips rather than with the chisel point. It also creates a chamfer at the edge of the drilled hole.

Since most drills have an included angle of 118°, when used as a spot, its depth should be $0.3D$, [Fig. 10.2(a)], where D is the diameter of the drill. If a 90° included angle spot drill is used, the depth of the spot drill will be $0.5D$ [Fig. 10.2(b)].

In addition to the above, the down movement of the drill needs to be programmed so that it feeds down *rapidly* to within a safe distance of the workpiece. Thereafter, the feed rate changes to the recommended feed for the work and drill materials that is used. For deep holes, or materials that are difficult to drill, it may be required that a "pecking" procedure be employed. This must be programmed into the drilling procedure. Once a hole is completed, the drill retracts at a rapid feed.

Example 1

(a) Use a $\frac{1}{2}$-in.-diameter drill with a 118° included angle to spot each hole in Fig. 10.1. Calculate the depth of the spot.

(b) Analyze and tabulate the movement of the drill.

Linear Interpolation: Drill Routines   Chap. 10

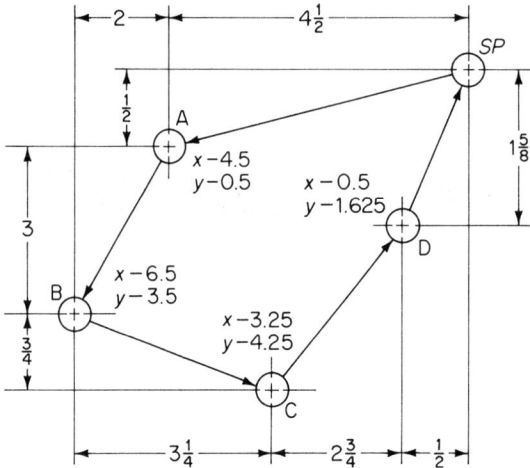

Figure 10.1

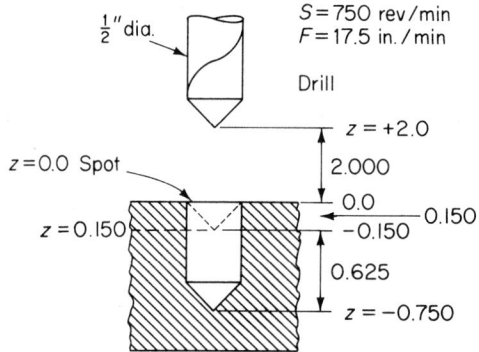

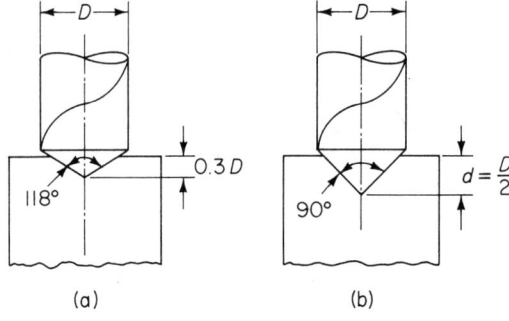

Figure 10.2

Sec. 10.1   Spot Drill   147

(c) Program Fig. 10.1 if you are to drill the four $\frac{1}{2}$-in. holes to a depth of $\frac{5}{8}$ in. The spindle speed is to be 750 rev/min and the feed is to be 17.5 in./min.

Solution (a) Since the $\frac{1}{2}$-in.-diameter drill bit has a 118° included angle, the drill should create a spot with a depth of

$$d = 0.300 \times 0.500 = 0.150 \text{ in.}$$

This is shown in Fig. 10.1.
(b) The analysis is as follows:
At SP:

$$X = 0.000$$
$$Y = 0.000$$

SP to A:

$$X = 0.000 - 4.500 = -4.500$$
$$Y = 0.000 - 0.500 = -0.500$$

A to B:

$$X = -4.500 - 2.000 = -6.500 \text{ in.}$$
$$Y = -0.500 - 3.000 = -3.500 \text{ in.}$$

B to C:

$$X = -6.500 + 3.250 = -3.250$$
$$Y = -3.500 - 0.750 = -4.250$$

C to D:

$$X = -3.250 + 2.750 = -0.500 \text{ in.}$$
$$Y = -4.250 + 2.625 = -1.625 \text{ in.}$$

D to SP:

$$X = 0.500 - 0.500 = 0.000$$
$$Y = -1.625 + 1.625 = 0.000$$

The tabulation is shown in Table 10.1(a).
(c) The program for Fig. 10.1 is shown in Table 10.1(b).

The G81 code programs a routine that permits drilling all four holes without repeating the approach and the spotting for each hole. In line N20 the X and Y movements (X-4.5, Y-0.5) are repeated in line N40. The depth of the hole (0.775) is programmed with a routine R value of 0.15. The R word directs the machine to rapid to the value given after the depth of the hole is rough drilled. In this case 17.5 in./min. for a depth of $\frac{5}{8}$ in.

TABLE 10.1(a)

X	Y	Z	Comment
0.000	0.000		SP
-4.500	-0.500		SP to A
Spot and drill			At A
-6.500	-3.500	-0.775	A to B
Spot and drill			At B
-3.250	-4.250	-0.775	B to C
Spot and drill			At C
-0.500	-1.625	-0.775	C to D
Spot and drill			At D
0.000	0.000		D to SP

TABLE 10.1(b)

Address	G Code	X Axis	Y Axis	Z Axis	R" Routine	Spindle Speed (rev/min)	Feed (ft/min)	M Function	Comment
N10	G00	-4.5	-0.5			750		M03	Spindle on: CW
N20									SP to A
N30									
N40	G81	-4.5	-0.5	.15	R.15		17.5		Drill A
N50		-6.5	-3.5	-0.775					Drill B
N60		-3.25	-4.25						Drill C
N70		-0.5	-1.625						Drill D
N80	G80								Cancel G81
N90	G00	0.0	0.0	2.0					Home
N100									D to SP
N110								M02	(end of program)

Sec. 10.1 Spot Drill

In N30 the drill rapids to 0.150 above Z zero. At this point the 1.75 in./min feed takes over. Each hole is drilled. After the last hole is drilled, the G80 code cancels the G81. The drill returns to the home position and the start point.

Example 2

(a) Using Fig. 10.3(a), calculate the movement of a 1-in. drill, 118° included angle. The feed is 30 in./min. The spindle speed is 700 rev/min. The hole is 1.500 in. deep. Make a drawing of the part and insert the X, Y, and Z values.

(b) Construct a table and insert these values.

(c) Write the program in the absolute mode.

Solution (a) The movement is:

$\oplus$ to SP: X = 0.000 − 2.000 = −2.000
        Y = 0.000 + 1.000 = 1.000
  Drill Z rapid to 0.000:
    Spot: at feed rate to −0.650
    Drill: at feed rate to −0.650 − 0.750 = −1.400
    Retract: −1.400 + 5.400 = 4.000
SP to A: X = −2.000 − 1.500 = −3.500
        Y = 1.000 + 1.500 = 2.500
  Drill: Repeat Z to 0.000
A to B: X = −3.500 − 3.000 = −6.500
        Y = 2.500   = 2.500
  Drill: Repeat Z to 0.000
B to C: X = −6.500 + 1.5 = −5.000
        Y = 2.500 + 2.598 = 5.098
  Drill: Repeat Z to 0.000
C to D: X = −5.000 − 2.500 = −7.500
        Y = 5.098 + 3.500 = 8.598
  Drill: Repeat Z to 0.000
D to E: X = −7.500 − 4.000 = −11.500
        Y = 8.598
  Drill: Repeat Z to 0.000
E to F: X = −11.500 − 3.000 = −14.500
        Y = 8.598 − 4.500 = 4.098
  Drill: Repeat Z to 0.000
F to $\oplus$: X = −14.500 + (3.000 + 4.000 + 2.500 + 1.500 + 1.500) − 2.000
        = 0.000
        Y = 4.098 − 3.098 − 1.000 = 0.000

Figure 10.3(b) shows the XY values.

(b) Table 10.2(a) shows these movements.

(c) Table 10.2(b) shows the program for Fig. 10.3(a).

150 Linear Interpolation: Drill Routines Chap. 10

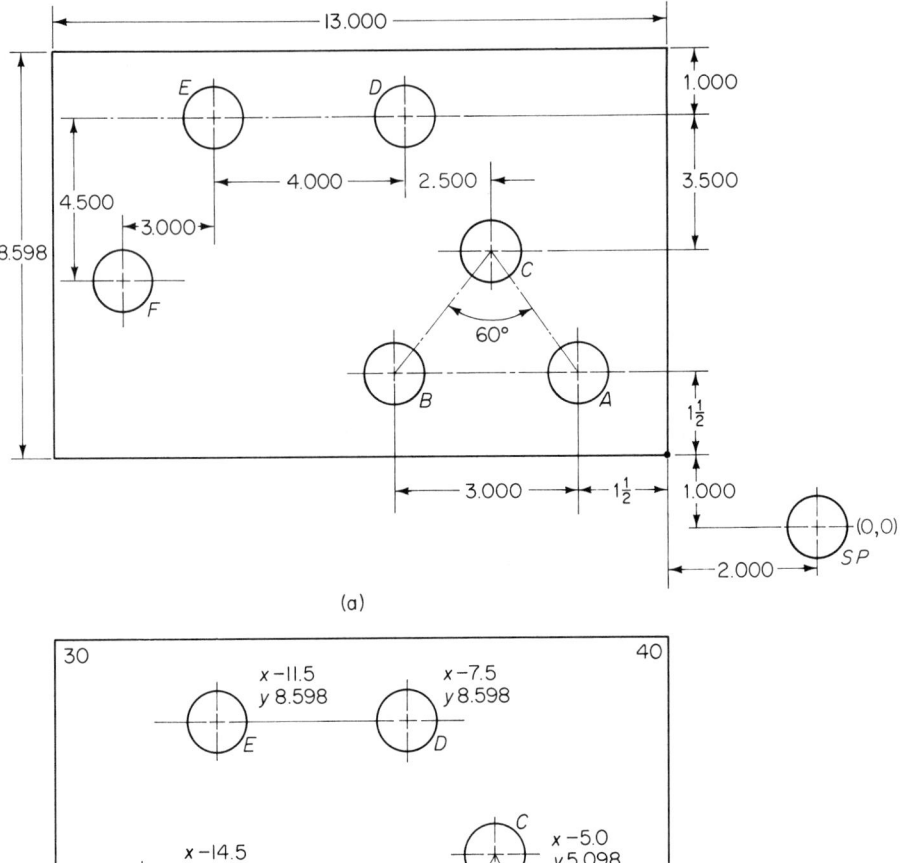

Figure 10.3

TABLE 10.2(a)

X	Y	Z	Move
−3.500	2.500		SP to A
		0.050	At A
		−1.300	At A
−6.500	2.500		At B
−5.000	5.098		At C
−7.500	8.598		At D
−11.500	8.598		At E
−14.500	4.098		At F
		2.000	At F
0.000	0.000		F to SP

TABLE 10.2(b)

N Address	G Code	X Axis	Y Axis	Z Axis	R Routine	Spindle Speed (rev/min)	F Feed	M Function	Comment
N10		0.0	0.0			700		M03	Spindle on: CW
N20	G00	−3.5	2.5						SP to A
N30				0.05					Rapid at A
N40	G81	−3.5	2.5	−1.3	0.05		F30		Drill A
N50		−6.5	2.5						Drill B
N60		−5.0	5.098						Drill C
N70		−7.5	8.598						Drill D
N80		−11.5	8.598						Drill E
N90		−14.5	4.098						Drill F
N100	G80								Cancel G81
N110	G00								Home
N120		0.0	0.0						F to SP
N130								M02	(end of program)

QUESTIONS AND PROBLEMS

Do not consider radius compensation in these problems.

10.1. Figure 10.4 is to be programmed in the absolute mode:
  (a) Calculate the point-to-point movements of the tool.
  (b) Insert the movements of part (a) into a drawing of the part.
  (c) Write the program

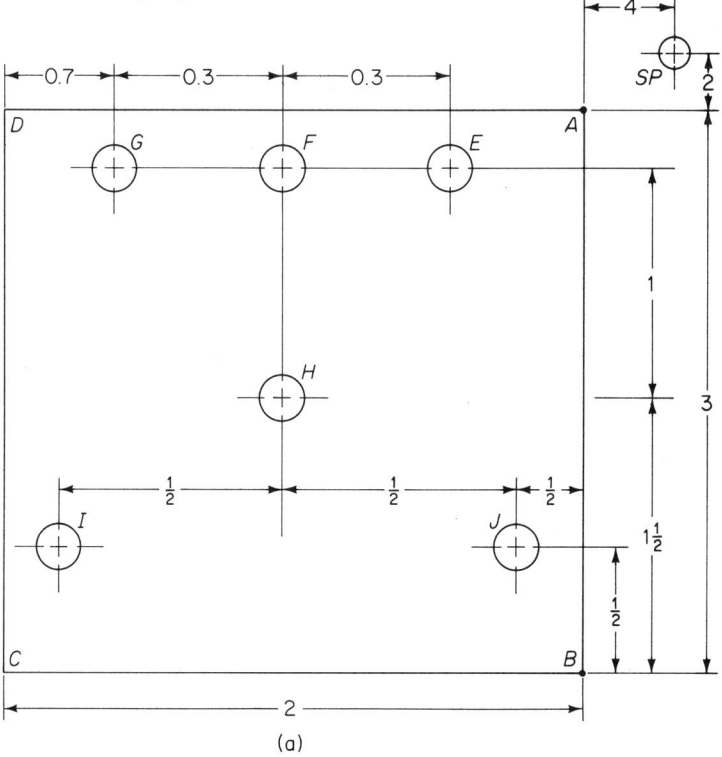

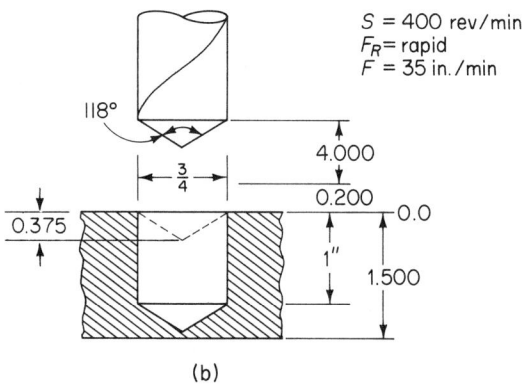

Figure 10.4

Chap. 10 Questions and Problems

10.2. Assume a ¾-in.-diameter drill with a 118° point. You are to drill holes E through I to a depth of 1.000 in. The spindle speed is 400 rev/min as shown in Fig. 10.4(a) and (b). The approach is 4.000 in. at a rapid feed, at which point a feed of 35 in./min takes over. Program the movement of the drill during the drilling operation. Remember that you must spot the hole before drilling it.

10.3. (a) Analyze the movement of the tool in Fig. 10.5(a) in the absolute mode.

(b) Analyze the drill spot and drill operation in Fig. 10.5(b).

(c) Combine parts (a) and (b) into a program for Fig. 10.5(a).

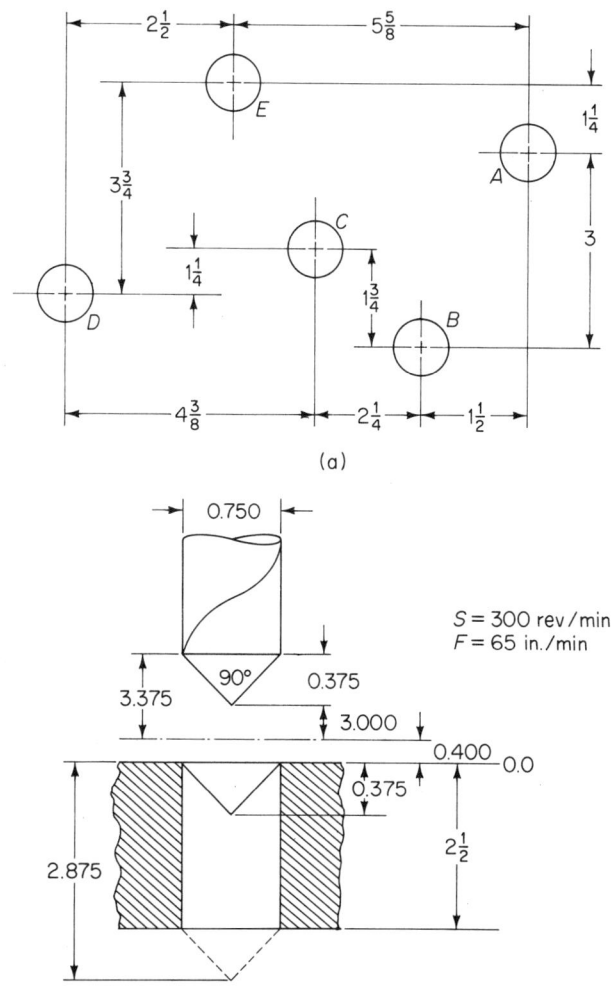

Figure 10.5

10.4. Program Fig. 10.6 in the absolute mode. Assume a 90° drill point on a $\frac{3}{4}$-in.-diameter drill. The hole is to be $\frac{5}{8}$ in. deep. Use a spindle speed of 350 rev/min and a feed of 25 in./min. Use a 0.300-in. approach to the top of the work. The latter does not include the depth of the spot.

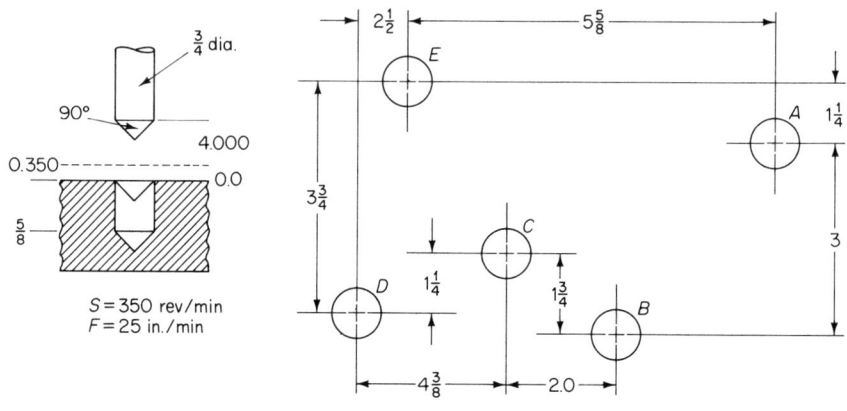

Figure 10.6

10.5. Program Fig. 10.7 in the absolute mode. Use a feed of 45 ft/min and a spindle speed of 250 rev/min. Use the drill and dimensions shown in Fig. 10.5(b).

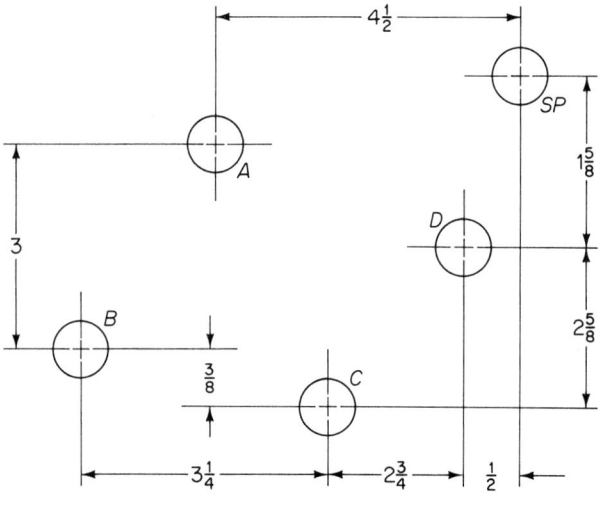

Figure 10.7

Chap. 10   Questions and Problems   155

10.6. Repeat Prob. 10.5 for Fig. 10.8.

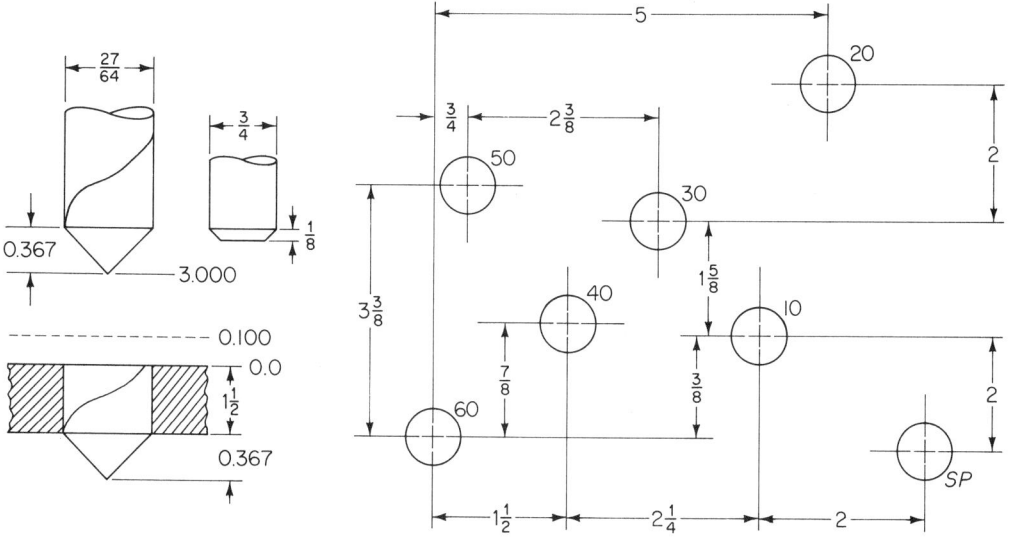

Figure 10.8

10.7. Program Fig. 10.9 in the absolute mode.

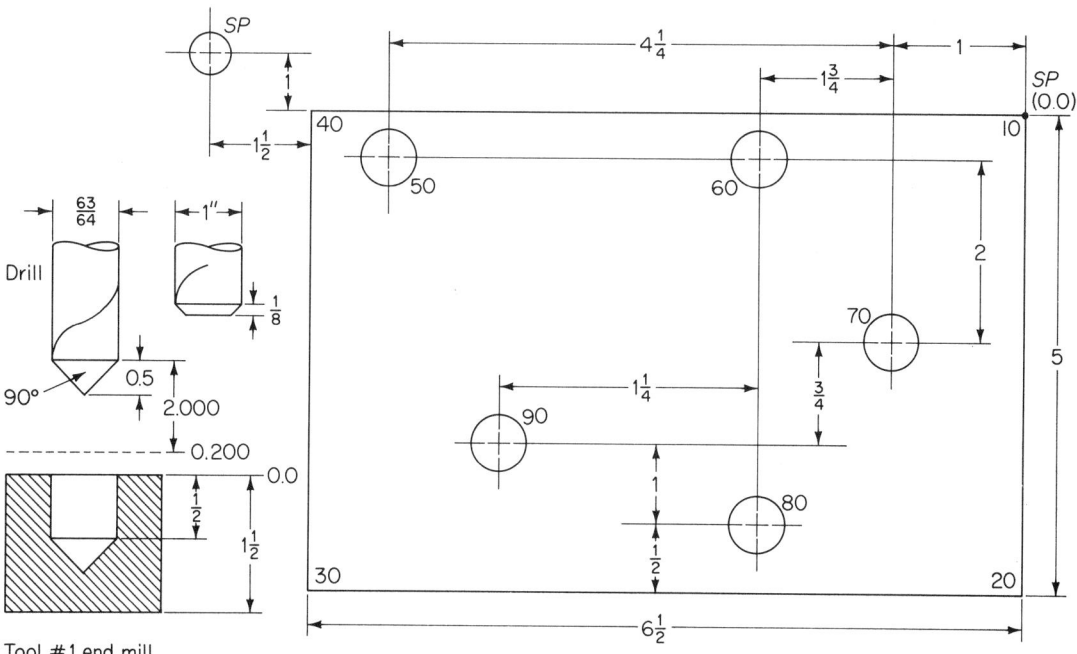

Tool #1 end mill
Tool #2 drill

Figure 10.9

10.8. Program Fig. 10.10 in the absolute mode. *Note:* The end mill is to be used to machine the hole and the slot.

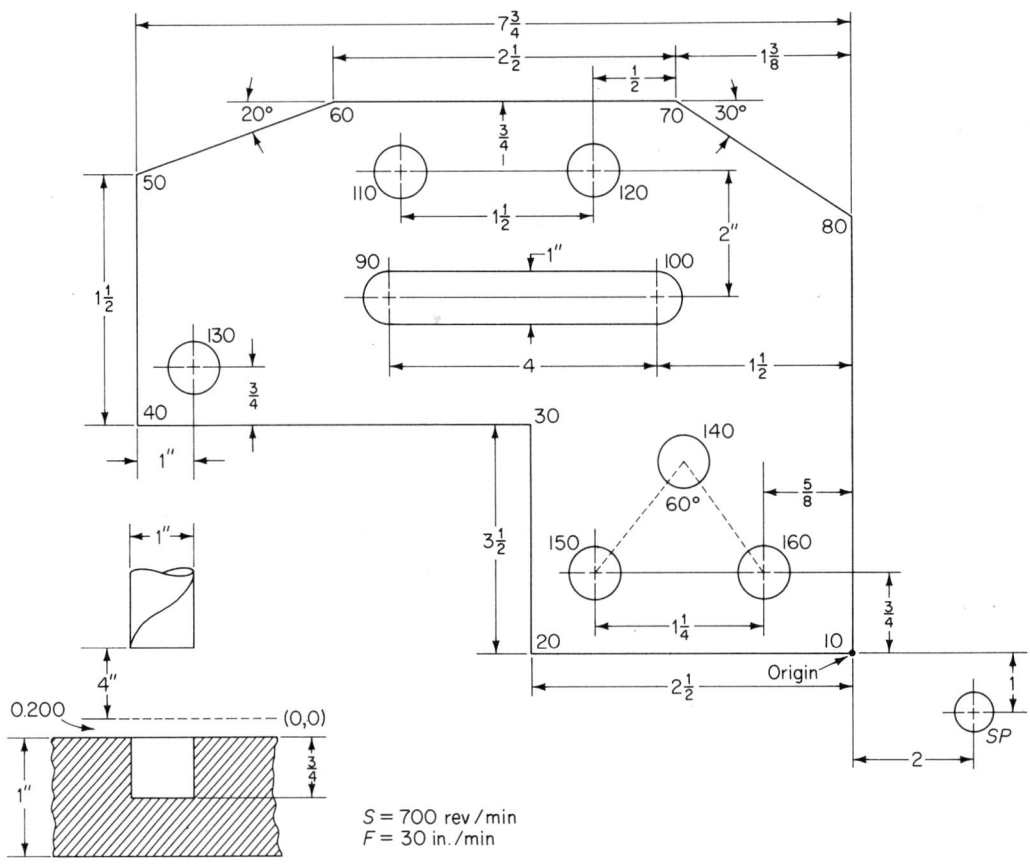

Figure 10.10

11

Tool Positioning and Threading

11.1 TOOL POSITIONING AND TOOL LENGTH OFFSET

Tools may be changed by using a quick-change tool holder or by using tools that have been stored in a tool changer. In either case the tools are preset. When using the quick-change tool holder, the tool lengths are set so that the difference in length between the shortest and the longest tool is calculated. The feed length of the shortest tool should not exceed the total stroke length of the quill.

Example 1

A $\frac{1}{16}$-in.-diameter drill, held in a collet, is to be exactly 1 in. shorter than a $\frac{1}{2}$-in end mill. This difference is to be accounted for in the programmed Z movement.

Solution

1. Set the *longest* tool ($\frac{1}{2}$-in. *end mill*) to any length in the tool holder.
2. Inset the tool holder into the spindle of the machine.
3. Jog the Z drive to the zero position at *midrange* of the quill travel. In Fig. 11.1(a) the midrange is shown as 3 in.
4. Bring the knee of the machine up so that the tool touches the surface of the work. The precise practice is to leave about $\frac{1}{16}$-in. clearance between the tool and the surface of the work, loosen the cutter clamp nut, and allow the tool to drop down

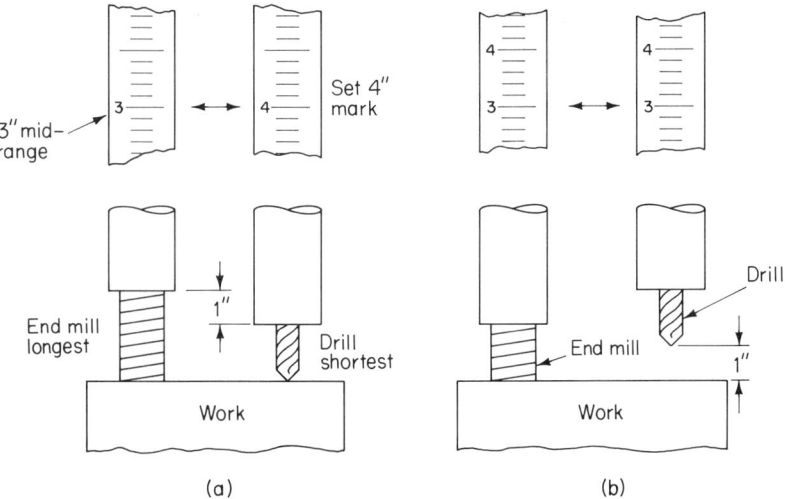

Figure 11.1

until it touches the surface of the work. Then the clamp nut is tightened. This setting becomes the reference for all additional tool settings.

5. Jog the quill up. Remove the holder and tool (end mill). Insert the holder and the drill with the shortest ($\frac{1}{16}$ in.) diameter.

6. Jog the quill down to a reading exactly 1 in. below the earlier setting [Fig. 11.1(a)]. In this case the quill is set at 4 in. Again, loosen the clamping nut, let the drill slide down until it contacts the surface of the work. Clamp the drill at this position in the holder.

7. The drill will be positioned exactly 1 in. less than the end mill [see Fig. 11.1(b)].

11.2 TOOL POSITIONING: TAPPING

Tapping with a CNC machine tool accomplished mainly with those that have automatic reversing spindles built into the canned cycle. The *lead* of the tap will control the *feed rate* and the rev/min rate of the tap. The end of the tap should be positioned about 1 in. above the work (Check the operating and code manual for the machine you are using.) There are several threads at the lead end of the tap that are incomplete. If a 100% finished thread-lead is required, the tap

Sec. 11.2 Tool Positioning: Tapping

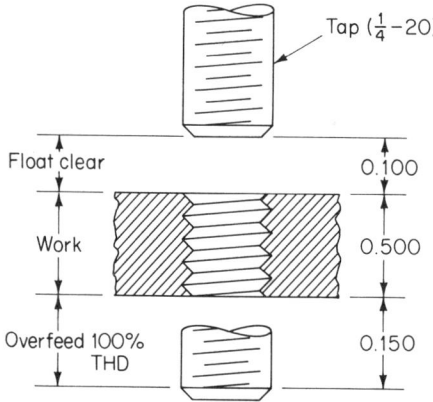

Figure 11.2

must be allowed to overtravel the hole. In Fig. 11.2 the lead of the tap is shown to be 0.100 in. and the overtravel is shown to be 0.150 in. It is recommended that the down feed of a tap be 95% the calculated feed rate. The following shows the method used to calculate the rpm, lead and feed required.

Revolutions per Minute

The approximate rev/min value may be calculated using the formula

$$R = 4 \times \frac{S}{M}$$

R = rev/min
M = major diameter tap, in.
S = speed, surface ft/min

For taps up to $\frac{1}{2}$ in., the recommended rev/min rate is 700. For larger-diameter taps, the formula should be used.

Lead

Another calculation is the *lead* of the tap. For single-lead threads, this turns out to be the reciprocal of the number of threads per inch. Thus

$$l = \frac{1}{T}$$

l = lead, in.
T = threads/in.

Feed

Once the revolutions per minute and the lead have been determined, the feed may be calculated as

$$F = l \times R$$

F = feed, in./min
l = lead, in.
R = rev/min

It is recommended that the *down feed* be 95% of the calculated F.

Example 2

A workpiece (Fig. 11.2) is $\frac{1}{2}$ in. thick. It is required to tap a $\frac{1}{4}$–20 thread into the workpiece. It is necessary that the tap have an entry clearance of 0.100 in. and an overfeed of 0.150 in. to produce a 100% completed thread. Assume the feed to be 44 in./min. Calculate:
(a) The rev/min.
(b) The lead.
(c) 100% of the feed rate.
(d) 95% of the feed rate.

Solution (a) The revolutions per minute for the tap is

$$R = \frac{4 \times S}{M} = \frac{4 \times 44}{\frac{1}{4}}$$

$$= 704 \text{ rev/min}$$

(b) The lead of this tap is

$$l = \frac{1}{20}$$

$$= 0.050 \text{ in.}$$

(c) The feed rate at 100% is

$$F_u = l \times R = 0.050 \times 704$$

$$= 35 \text{ in./min}$$

(d) At 95%, the feed is

$$F_d = F_u \times 0.95\% = 35 \times 0.95$$

$$= 33 \text{ in./min}$$

TABLE 11.1

N Address	G Code	X Axis	Y Axis	Z Axis	R Routine	Spindle Speed (rev/min)	Feed (in./min)	T/H Tool	M Function	Comment
N10	G00G90	0.0	0.0	2.0				T01/H01	M06	29/64 drill
N20						500			M03	Spindle on
N30										Rapid to 0.1
N40	G81	0.0	0.0	-0.65	0.1		20.0			Drill hole
N50	G80									Cancel drilling cycle
N60	G00	0.0	0.0	2.0				T02/H02	M06	¼-20 tap
N70						700			M03	Spindle on
N80										Rapid to 0.1
N90	G84	0.0	0.0	-0.65	0.1		33.0			Tap hole
N100	G80									Cancel tap cycle
N110	G00				2.0					Spindle home
N120									M02	(end of program)

Example 3

Write the program to produce the thread in Example 2. Use a 29/64 in (0.453) tap drill at a spindle speed of 500 rev/min.
(a) Analyze the sequence of operations necessary to tap the hole.
(b) Program the sequence.

Solution (a) The analysis is as follows:

> Start program
> Drill cycle: 500 rev/min, feed 20 in./min
> Spindle on 704 rev/min
> Enter tapping cycle: spindle speed 700 rev/min
> Cancel tapping cycle
> End of program

(b) The program is shown in Table 11.1.

11.3 PROGRAMMABLE CYCLE FILES

Many operations that are repetitive may be programmed and stored in a "file." These operations may be called up with a G code. This saves time when programming and also shortens the length of the program. Files are discussed more fully in Chapter 15. Before attempting a file program, a preliminary tool analysis should be made. The tool analysis table (Table 11.2) should be done first, followed by the file program.

Example 4

Figure 11.3(a) is to be programmed in the absolute mode.

(a) Do a tool analysis.

TABLE 11.2

G Code	T Tool	Tool	Spindle Speed (rev/min)	Operation	Feed (in./min)
80		Cancels canned cycle			
81	T02	⅛ drill	1500	Drill	15
	T03	No. 3 drill	1375	Drill	35
	T04	0.238 drill	1000	Drill	12.5
84	T05	¼-28 tap	700	Tap	25
85	T06	¼ reamer	900	Reamer	10

Sec. 11.3   Programmable Cycle Files

TABLE 11.3

T Tool	G Code	Operation	F	Z
	80	Cancel operation		
T02	81	Drill	15.0	-0.450
T03	81	Drill	35.0	-0.500
T04	81	Drill	12.5	-0.500
T05	84	Tap	25.0	-0.600
T06	85	Ream	10.0	-0.400

(b) Do a file program.

(c) Calculate the point-to-point movement of the tool.

(d) Using a cycle file, write the program. Assume that the outside of the part is the center of an end mill. Write the program for the outside of the part.

Solution   (a) The tool analysis is shown in Table 11.2. The tap feed may be determined by multiplying the rev/min value by the reciprocal of the number of threads per inch (the pitch). In this case the feed will be

$$F = R\left(\frac{1}{T}\right) = 700\left(\frac{1}{28}\right) = 25 \text{ in./min}$$

(b) The program in the file is shown in Table 11.3.

(c) The movements of the tool are shown in Fig. 11.3(b). They are:

At the tool change position:

$$X = 4.000$$
$$Y = 2.000$$

From ⊕ to A:

X = -4.000 + 4.000 = 0.000

Y = -2.000 + 2.000 = 0.000

From A to B:

X =   0.000

Y = -0.300

From F to G:

X = -1.000 - 0.300 = -1.300

Y = -0.500

From G to H:

X = -1.300 + 0.300 = -1.000

Y = -0.500 - 1.000 = -1.500

TABLE 11.4(a)

Address N	Code G	Axis X	Axis Y	Axis Z	Routine R	Spindle Speed (rev/min)	Feed (in/min)	Tool T/H	Misc. M	Comment
N10	G00	1.0	2.0	4.0				T01/H01	M06	At SP
N20	G90									At SP
N30						1200			M03	At SP
N40		0.0	0.0	0.1						SP to A
N50	G01	0.0	-3.0	-0.25			15.0			A to B
N60		-2.0								B to C
N70			0.0							C to D
N80		0.0								D to A
N90	G00	1.0	2.0	4.0						A to SP
N120		-0.7	-0.5	0.1						SP to E
N130	G81	-0.7	-0.5	-0.45	R0.1		15.0			At E
N140		-1.0	-0.5							E to F
N150		-1.3	-0.5							F to G
N160	G80									Cancel G81
N170	G00	1.0	2.0							G to SP
N180								T03/H03	M06	At SP
N190						1375			M03	At SP
N200		-1.0	-1.5	-1.0						SP to H
N210	G81	-1.0	-1.5	-0.5	R0.1		35.0			At H

```
N220        G80                                                              Cancel
N230        G00         1.0     2.0                                          G81
N240                                                                         H to SP
N250                                                     T04/H04  M06        At SP
N260                                                              M03        At SP
N270        G81        -1.5    -2.5     0.1    R.05     1000      12.5       SP to I
N280                   -1.5    -2.5    -0.5                                  At I
N290        G80        -0.5    -2.5                                          I to J
N300                                                                         Cancel
N310        G00         1.0     2.0                                          G81
N320                                                     T05/H05  M06        J to SP
N330                                                              M03        At SP
N340        G84        -1.0    -1.5     0.1    R0.1      700      25.0       At SP
N350        G80        -1.0    -1.5    -0.6                                  SP to H
                                                                             At H
                                                                             Cancel
N360        G00         1.0     2.0                                          G84
N370                                                     T06/H06  M06        H to SP
N380                                                              M03        At SP
N390        G85        -1.5    -2.5    -0.4    R1.0      900      10.0       At SP
N400                   -1.5    -2.5                                          SP to I
N410                   -0.5    -2.5                                          At I
N420        G80                                                              I to J
                                                                             Cancel
N430        G00         1.0     2.0                                          G85
N440                    1.0     2.0     4.0                       M02        J to SP
                                                                             EOP
```

TABLE 11.4(b)

Lines	Operation
N10	This line places the tool at the tool load position. The G00 code calls for a rapid feed.
N20	H01 is the command that sets the tool length offset (TLO) for tool number 1. M06 loads the tool and enters the value of H01.
N30	The block sets the feeds and speeds. The M03 starts the spindle.
N40	Moves the spindle to point A, feeds the spindle to the 0.100 in depth (assuming the center of the cutter).
N50, N60, N70, N80	Trace the outline of the workpiece. The cutter has returned to point A.
N90	The G00 command sends the tool to the tool change position.
N100	Repeats line N20 for tool number 2.
N110	Starts the spindle.
N120	Moves the spindle to E, 0.100 in. above the work.
N130	The G81 sets the drilling routine and drills the hole to a depth of -0.450 in. When the tool has completed its cut, it rapid feeds to the R plane, 0.100 in. above the work. The feed is 15 in./min.
N140	G81 has not been canceled. Therefore, the drill will move to hole F and repeat the N130 sequence.
N150	The hole at G will also be drilled.
N160	G80 cancels the drilling routine.
N170	G00 sends the tool back to the tool change position.
N180	Loads the tool and offset for tool number 3.
N190	M03 starts the spindle and sets the spindle speed at 1375 rev/min.
N200	Moves the spindle to position H, 0.100 in. above the work.

N210	Drills hole H 0.500 in. deep at 35 in./min.
N220	G80 cancels the drill routine.
N230	G00 sends the tool back to the tool change position.
N240	Repeats line N20 for tool number 4.
N250	M03 starts the spindle rotating at 1000 rev/min.
N260	Moves the spindle to hole I.
N270	Drills hole I -0.500 in. deep. G81 sets the drill routine.
N280	Drills hole J -0.500 in. deep.
N290	G80 cancels the drill routine.
N300	G00 sends the tool back to the tool change position.
N310	Loads tool and offset for tool number 5.
N320	M03 starts spindle at 700 rev/min.
N330	Moves the tool to position H 0.100 in. above the work.
N340	G84 sets the tap cycle. When the tap reaches a depth of 0.600 in., the spindle will reverse automatically. When the spindle reaches the R 0.100 in. plane, the spindle will automatically reverse to its normal direction of rotation.
N350	G80 cancels the tap cycle.
N360	G00 sends the spindle to the tool change position.
N370	Loads tool and offset for tool number 6.
N380	M03 starts spindle rotation at 900 rev/min.
N390	Moves the tool to hole I, 0.100 in. above the work.
N400	G85 will cause the reamer to feed at a rate of 10 in./min to a depth of -0.400 in. The reamer will then retract to the R plane 0.100 in. above the work.
N410	Hole J will be reamed.
N420	G80 cancels the G85 reaming operation.
N430	G00 causes the tool to move to the tool change position.
N440	M02 ends the program (EOP).

From B to C:

X = −0.200

Y = −3.000

From C to D:

X = −2.000

Y = −3.000 + 3.000 = 0.000

From D to A:

X = −2.000 + 2.000 = 0.000

Y = 0.000

From A to E:

X = 0.000 − 0.700 = −0.700

Y = 0.000 − 0.500 = −0.500

From E to F:

X = −0.700 − 0.300 = −1.000

Y = −0.500

From H to I:

X = −1.000 − 0.500 = −1.500

Y = −1.500 − 1.000 = −2.500

From I to J:

X = −1.500 + 1.000 = −0.500

Y = −2.500

From J to A:

X = −0.500 − 0.500 = 0.000

Y = −2.500 + 2.500 = 0.000

From A to ⊕:

X = 0.000 + 4.000 = 4.000

Y = 0.000 + 2.000 = 2.000

(d) The program is shown in Table 11.4(a).

Table 11.4(b) explains each of the lines in the program, Table 11.4(a). The student should correlate the explanations with each address in the program.

QUESTIONS AND PROBLEMS

11.1. Explain the procedure for setting the tool length offset when the longest tool is $1\frac{1}{2}$ in. longer than the shortest tool.

11.2. Given a $\frac{3}{4}$-10 thread with a lead of $\frac{1}{4}$ in. and a required overtravel of 0.060 in. Calculate:
 (a) The rev/min value.
 (b) The lead of the tap.
 (c) The 100% feed.
 (d) The 95% feed.

11.3. What purpose does a cycle file serve?

11.4. Program Fig. 11.3 in the absolute mode. Use a canned cycle file. Use a 1" end mill to machine the outside of the workpiece. Disregard radius offset.

Chap. 11 Questions and Problems

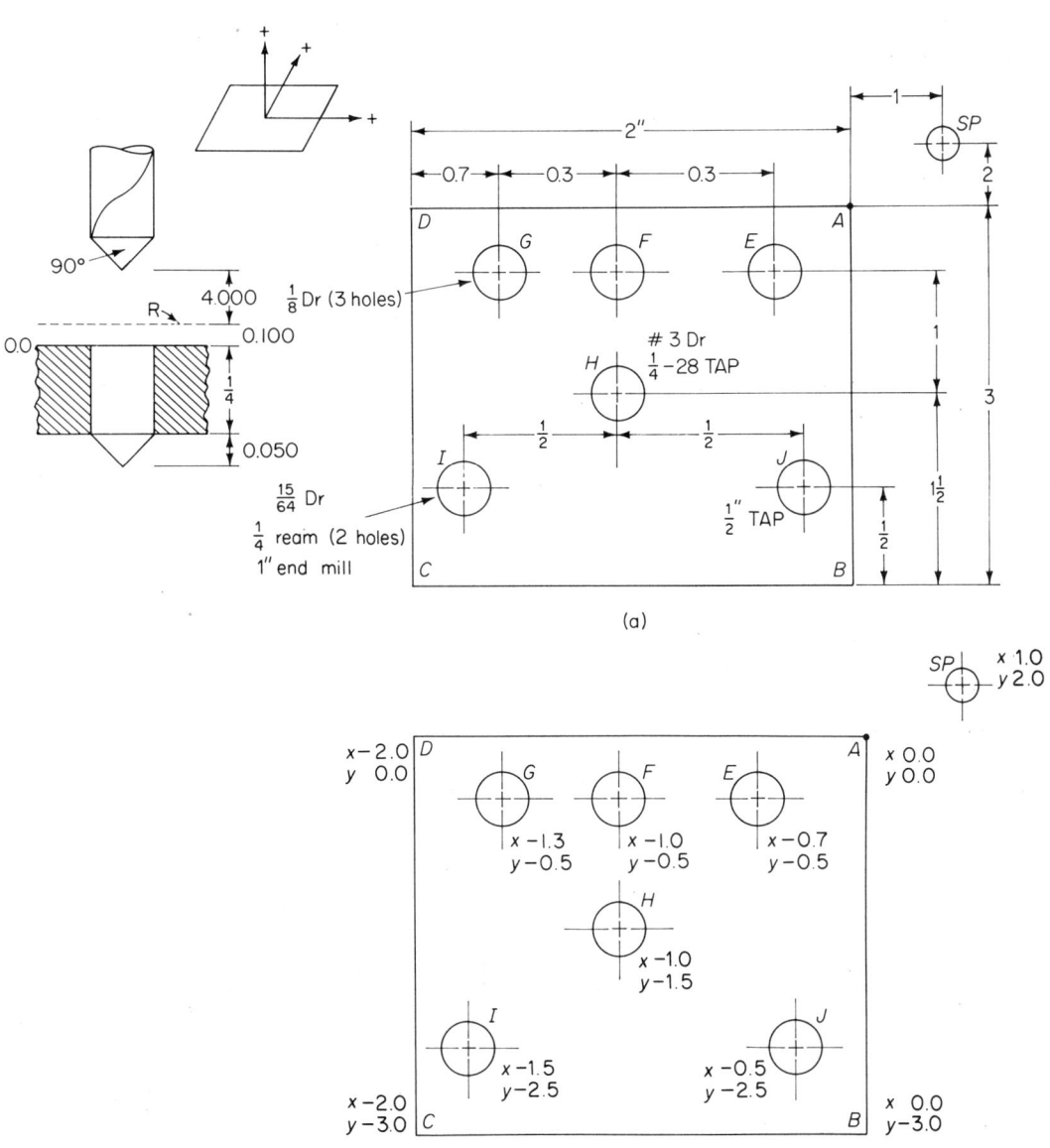

Figure 11.3

170   Tool Positioning and Threading   Chap. 11

11.5. Repeat Prob. 11.4 for Fig. 11.4.

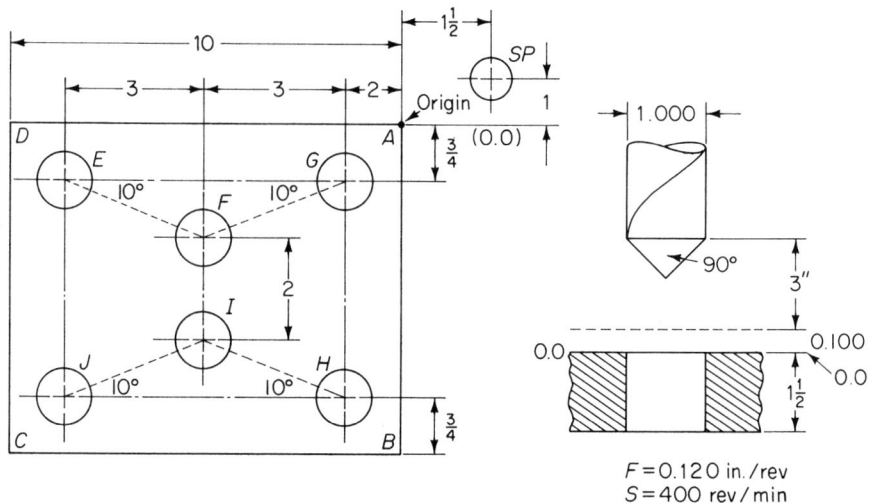

Figure 11.4

11.6. Repeat Prob. 11.4 for Fig. 11.5.

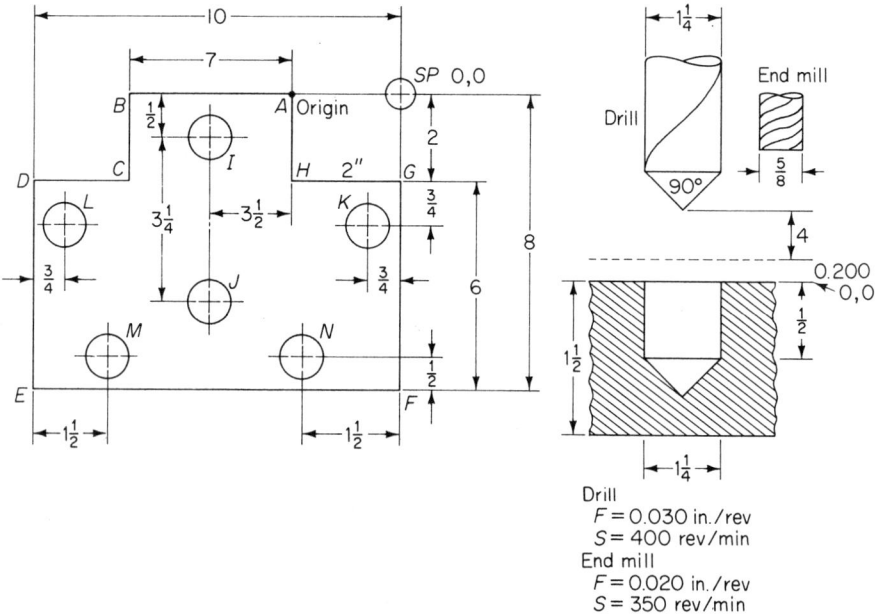

Figure 11.5

Chap. 11 Questions and Problems

11.7. Repeat Prob. 11.4 for Fig. 11.6.

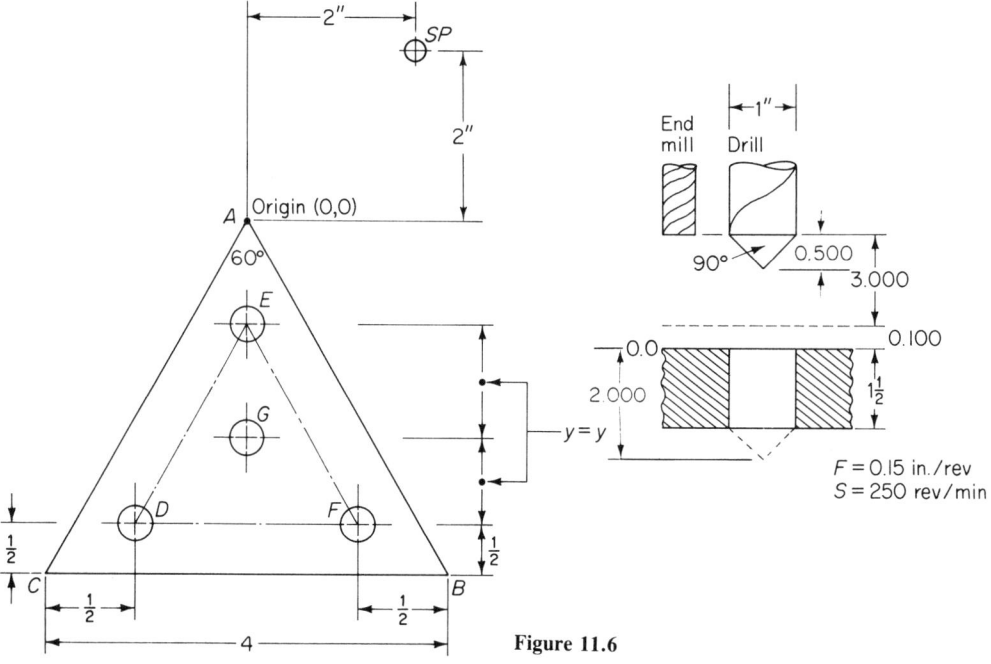

Figure 11.6

11.8. Repeat Prob. 11.4 for Fig. 10.10.

11.9. With reference to Fig. 11.3(a), assume that holes E, F, and G are drilled and tapped for a $\frac{3}{4}$-10 thread. J, I, and H are drilled with a $\frac{1}{2}$-in.-diameter drill. A 90° countersink is used to generate a $\frac{7}{8}$-in.-diameter chamfer.
(a) Program the $\frac{3}{4}$-10 thread (E, F, G).
(b) Program the drilled holes (J, I, H).

11.10. Use a canned cycle for Fig. 10.8. Program the seven holes. Holes of 10, 20, and 30 are drilled and reamed with a $\frac{47}{64}$ drill (90° included angle) and a $\frac{3}{4}$-in.-diameter reamer ($\frac{1}{8}$ in. lead), respectively. Holes 40, 50, and 60 are threaded with a $\frac{1}{2}$-13 tap. The tap has a $\frac{1}{4}$-in. lead.

11.11. Repeat Prob. 11.10 for Fig. 10.9. Use a $\frac{63}{64}$ in. drill and a 1 in. reamer for holes 50, 60, 70 and a $\frac{21}{32}$ in. drill and a $\frac{1}{2}$-13 tap for holes 80 and 90.

11.12. Repeat Prob. 11.10 for Fig. 10.10. The three holes, 140, 150, and 160, are to be drilled and reamed. Holes 110, 120, and 130 are to be drilled through and tapped with a $\frac{5}{8}$-18 tap. The tap has a $\frac{3}{8}$ in. lead. This drill point is 90°.

11.13. Using a canned cycle file, program Fig. 10.3(a). Holes A, B, and C are drilled and reamed to 1.000 in. Holes 40, 50 and D, E, and F are to be tapped with a $\frac{1}{2}$-20 tap; lead $\frac{3}{8}$ in., rev/min 80 ft/min. Drill, tap and ream through the workpiece.

12

Circular Interpolation: Absolute Mode

12.1 CIRCULAR INTERPOLATION

When G74 is activated for single quadrant circular interpolation, one line of information can define an arc in one quadrant only. Therefore no arc can be greater than 90°. Circular interpolation is accomplished with the code letters I, J and K. The I code is related to the X axis, the J code to the Y axis, and the K code to the Z axis. They are programmed without the use of the plus (+) and minus (−) signs.

In either the absolute, or the incremental mode (G74), I and J are incremental. The distances are taken from the center of the arc to the start point of the arc. The X, Y, and Z commands must carry the proper signs.

A G code must be used to define the plane through which the cutter moves. In addition, in this system, a G02 or G03 is used to indicate a clockwise or counterclockwise motion of the tool: G02 is clockwise, G03 is counterclockwise [see Fig. 12.1(a) and (b)].

The I and J distances are programmed along the X and Y axes [see Fig. 12.1(a) and (b)] and applied to the end of the arc as shown in Fig. 12.1(c). Thus

G02	X	Y	I	J	F
CW	Endpoints of arc		Arc center offsets		Feed rate

174 Circular Interpolation: Absolute Mode Chap. 12

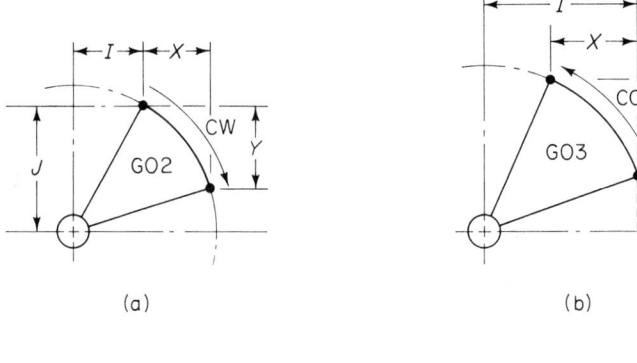

(a)   (b)

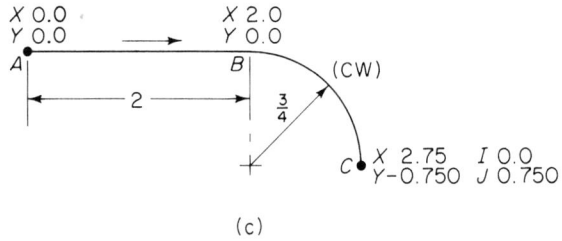

(c)

Figure 12.1

Example 1

Program Fig. 12.2(a).
  (a) Show the analysis at each point.
  (b) Insert the X, Y, I, J codes on a drawing [Fig. 12.2(b)].
  (c) Program the figure in the absolute mode.

Solution (a) The analysis is shown in Table 12.1.
  Note that the X and Y movements are absolute and signed, whereas the I and J movements are *incremental and unsigned*.
  (b) The X, Y, I, and J coded dimensions are shown in Fig. 12.2(b).
  (c) The program is as shown in Table 12.2.

Example 2

Figure 12.3(a) is to be programmed in the absolute mode.
  (a) Analyze and calculate the code movements *A* through *H*.
  (b) Insert the dimensions at each point in a drawing.
  (c) Write the program.

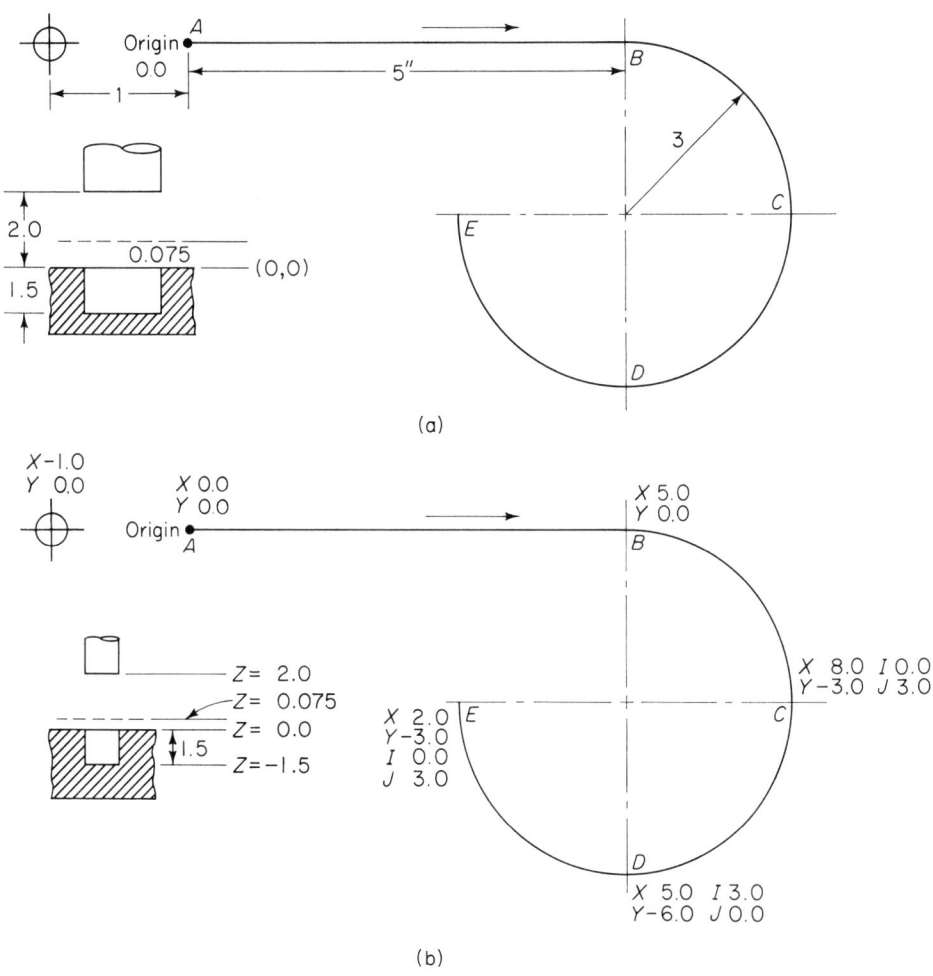

Figure 12.2

TABLE 12.1

Move	X	Y	Z	I	J
At SP	-1.000	0.000	2.000		
			0.075		
			-1.500		
SP to A	0.000				
A to B	5.000				
B to C	8.000	-3.000		0.000	3.000
C to D	5.000	-6.000		3.000	0.000
D to E	2.000	-3.000		0.000	3.000
At E			2.000		
E to SP	-1.000	0.000			

TABLE 12.2

N Address	G Code	X Axis	Y Axis	Z Axis	I Arc	J Arc	Feed (in./min)	Spindle Speed (rev/min)	T/H Tool	M Function	Comment
N10	G0G90	-1.0	0.0	2.0				2500	T01/H01	M06	At A
N20		0.0		0.075							Top work
N30	G01			-1.5			0.75				Hole depth
N40		5.0					1.5				A to B
N50	G02	8.0	-3.0		0.0	3.0					B to C
N60		5.0	-6.0		3.0	0.0					C to D
N70		2.0	-3.0		0.0	3.0					D to E
N80	G00			2.0							At E
N90		-1.0		0.0							E to SP
N100										M02	(end of program)

Sec. 12.1   Circular Interpolation   177

Solution   (a) The analysis from point-to-point generates the X and Y values at:

Origin to SP:

$X = -2.250 - 1.750 - 1.000 \qquad = -5.000$
$Y = -3.000 + 0.750 \qquad = -2.250$

SP to A:

$X = -5.000 + 1.000 + 1.750 \qquad = -2.250$
$Y = -2.250 - 0.750 \qquad = -3.000$

A to B:

$X = -2.250 + 5.750 \qquad = 3.500$
$Y = -3.000 \qquad = -3.000$

B to C:

$X = 3.500 + 0.500 \qquad = 4.000 \qquad I = 0.000$
$Y = -3.000 + 0.500 \qquad = -2.500 \qquad J = 0.500$

C to D:

$X = 4.000 \qquad = 4.000$
$Y = -2.500 + 2.500 \qquad = 0.000$

D to E:

$X = 4.000 - 3.000 \qquad = 1.000 \qquad I = 3.000$
$Y = 0.000 + 3.000 \qquad = 3.000 \qquad J = 0.000$

E to F:

$X = 1.000 - 3.000 \qquad = -2.000$
$Y = 3.000 \qquad = 3.000$

F to G:

$X = -2.000 - 2.000 \qquad = -4.000 \qquad I = 0.000$
$Y = 3.000 - 2.000 \qquad = 1.000 \qquad J = 2.000$

G to H:

$X = -4.000 \qquad = -4.000$
$Y = 1.000 - 2.250 \qquad = -1.250$

H to A:

$X = -4.000 + 1.750 \qquad = -2.250 \qquad I = 1.750$
$Y = -1.250 - 1.750 \qquad = -3.000 \qquad J = 0.000$

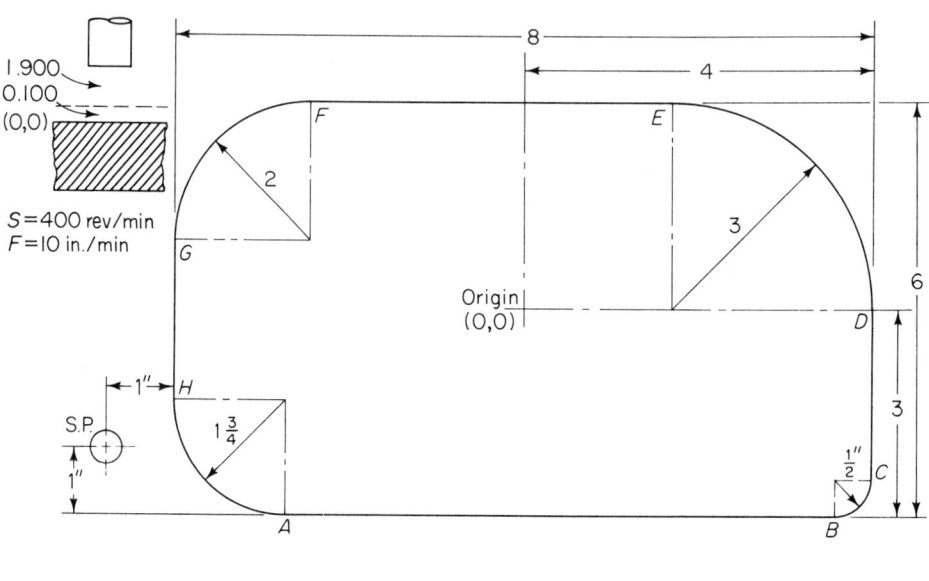

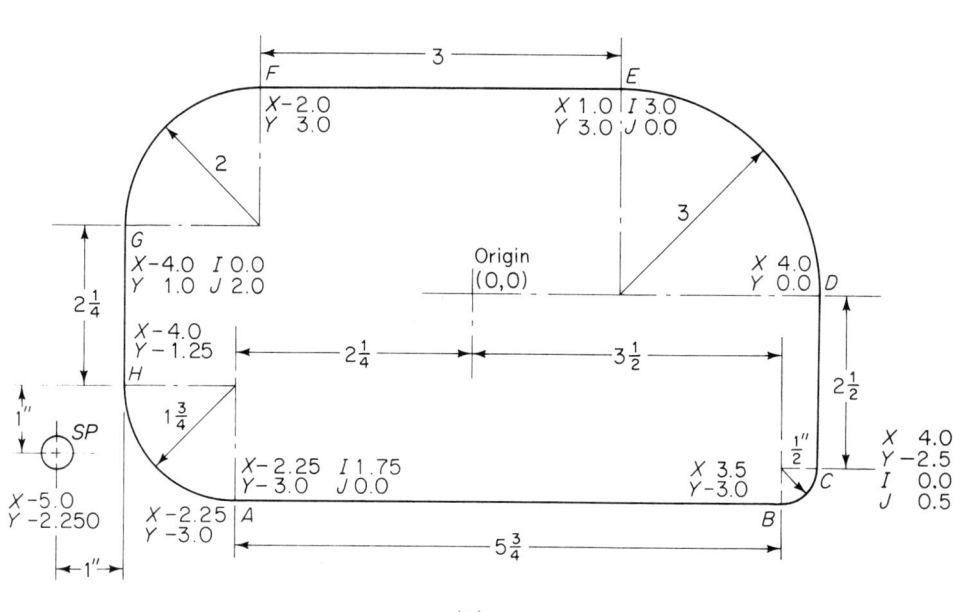

Figure 12.3

A to SP:

$X = -2.250 - 2.750 = -5.000$
$Y = -3.000 + 0.750 = -2.250$

(b) The dimensions have been inserted into Fig. 12.3(b).
(c) The program for Fig. 12.3(a) is as shown in Table 12.3.

TABLE 12.3

N Address	G Code	X Axis	Y Axis	Z Axis	I Arc	J Arc	Feed (in./min)	Spindle Speed (rev/min)	T/H Tool	M Function	Comment
N10	G00	-5.0	-2.25	2.0				400	T01/H01	M06	Rapid at SP
N20				0.1							At SP
N30		-2.25	-3.0								SP to A
N40	G01	3.5					10.0				A to B
N50	G74 G03	4.0	-2.5		0.0	0.5					B to C
N60	G01	4.0	0.0								C to D
N70	G03	1.0	3.0		3.0	0.0					D to E
N80	G01	-2.0									E to F
N90	G03	-4.0	1.0		0.0	2.0					F to G
N100	G01		-1.25								G to H
N110	G03	-2.25	-3.0		1.75	0.0					H to A
N120	G00			2.0							Rapid at A
N130		-5.0	-2.25								A to SP
N140										M02	(end of program)

12.2 MULTIQUADRANT CIRCULAR INTERPOLATION

Multiquadrant circular interpolation permits shorter programs to be written. It is *not* necessary to program at every 90° point in an arc. The G75 code may be used in either the absolute or incremental mode. When using the G75 code, all X, Y, Z, I, J, and K values, including the zeros, must be programmed. In addition, the G02 (clockwise) or G03 (counterclockwise) movements must be used.

The following codes must also be observed:

```
The center of the arcs are programmed as I and J.
I and J are unsigned.
G00 indicates a rapid feed mode.
G01 indicates a select feed mode.
G90 is the absolute mode.
G91 is the incremental mode.
G75 indicates multiquadrant circular interpolation.
G02 indicates clockwise circular interpolation.
G03 indicates counterclockwise circular interpolation.
```

There are several additional methods for dealing with circular interpolation. Each serves its own special purpose. To begin with, the linear X, Y, and Z are converted into circular I, J, and K movements. As we have seen, the I command is coupled with the X, the J with the Y, and the K with the Z command. In the G75 G90 absolute mode, the I, J, and K values are unsigned and are taken from the center of the arc to the start point of the arc. In the G75 G91 incremental mode, the I, J, and K values are signed and incremental. They are also taken from the center of the arc to the start point of the arc. Also, the I, J and K commands, although determined at the start of the arc, are programmed with the *X, Y, and Z commands at the end of the arc*. This is shown in Fig. 12.4.

Example 3

Program Fig. 12.4(a) using multiquadrant circular interpolation in:

(a) The absolute mode.
(b) The incremental mode.

Solution (a) In the *absolute* mode the program is as shown in Table 12.4(a) and Fig. 12.4(b).

(b) In the incremental mode the program is as shown in Table 12.4(b) and Fig. 12.4(c).

Sec. 12.2 Multiquadrant Circular Interpolation

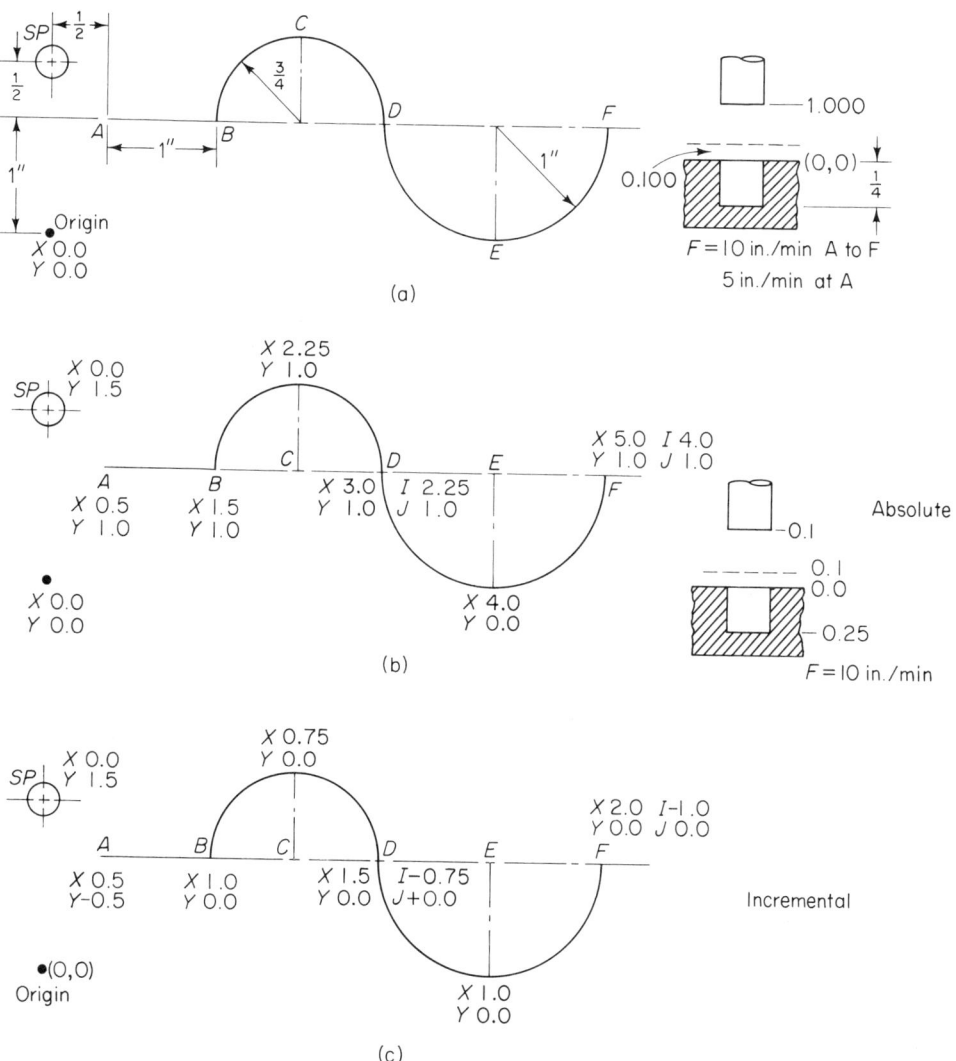

Figure 12.4

Example 4

Program the arc AB (Fig. 12.5) in a counterclockwise direction.

(a) Do the point-to-point analysis in the absolute mode.
(b) Write the program in the absolute mode (G75).
(b) Write the program in the incremental mode (G74).

TABLE 12.4(a)

N Address	G Code	X Axis	Y Axis	Z Axis	I Arc	J Arc	Feed (in./min)	Spindle Speed (rev/min)	T/H Tool	M Function	Comment
N10	G00G90	0.0	1.5	1.0					T01/H01	M06	Rapid at SP
N20	G75			0.1							At SP
N30		0.5	1.0								SP to A
N40	G01			-0.25							At A
N50		1.5					5.0				A to B
N60	G02	3.0			2.25	1.0	10.0				B to D
N70	G03	5.0			4.0	1.0					D to F
N80	G00			1.0							Rapid at F
N90		0.0	1.5								F to SP
N100										M02	(end of program)

TABLE 12.4(b)

N Address	G Code	X Axis	Y Axis	Z Axis	I Arc	J Arc	Feed (in./min)	Spindle Speed (rev/min)	T/H Tool	M Function	Comment
N10	G00G91	0.0	1.5	1.0					T01/H01	M06	Rapid at SP
N20	G75			-0.9	at A						
N30		0.5	-0.5								SP to A
N40	G01			-0.35							At A
N50		1.0	0.0				5.0				A to B
N60	G02	1.5	0.0		-0.75	0.0	10.0				B to D
N70	G03	2.0	0.0		-1.0	0.0					D to F
N80	G00			1.25							Rapid up
N90		-5.0	0.5								F to SP
N100										M02	(end of program)

Sec. 12.2   Multiquadrant Circular Interpolation

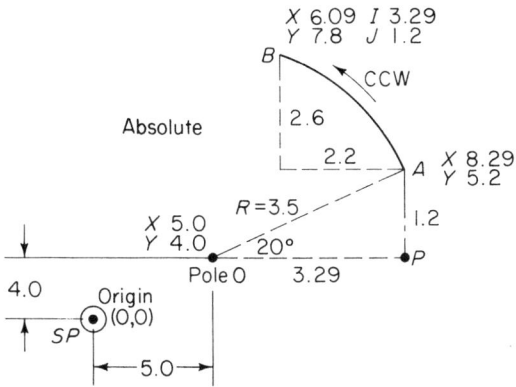

Figure 12.5

Solution (a) The point-to-point analysis in the absolute mode is as follows. In the X and Y directions in triangle OPA:

$$\cos 20° = \frac{OP}{3.500}$$

$$OP = 3.500 \times \cos 20°$$

$$= 3.290$$

$$\sin 20° = \frac{AP}{3.500}$$

$$AP = 3.500 \times \sin 20°$$

$$= 1.200$$

Therefore, in the absolute system:

At A:

$$X = 5.000 + 3.290 = 8.290$$

$$Y = 4.000 + 1.200 = 5.200$$

At B:

$$X = 8.290 - 2.200 = 6.090$$

$$Y = 5.200 + 2.600 = 7.800$$

(b) The program line (Fig. 12.5) in the *absolute* system would read

```
N10   G90   G01   X8.29   Y5.2              F150
N15   G75   G03   X6.09   Y7.8   I6.09   J5.2
```

Thus the G90 indicates the absolute mode, G01 the feed mode, G03 the counterclockwise arc direction, and I and J are unsigned arc offset from the origin.

(c) In the *incremental* system the program would be

N10   G91   G01   X8.29   Y5.2   F150
N15   G74   G03   X-2.2   Y2.6   I2.2   J0.

Note that the movement in the incremental system from A to B is simply the X or Y distance from A to B.

Example 5

Refer to Fig. 12.6:

(a) Do the analysis in the absolute mode.
(b) Write the program in the absolute mode.
(c) Write the program in the incremental mode.

Solution (a) The analysis in the absolute mode is as follows.
Starting at A:

$$X = 0.000 + 4.000 = 4.000$$
$$Y = 0.000 = 0.000$$

A to B:

$$X = 4.000 + 3.000 = 7.000$$
$$Y = 0.000 - 3.000 = -3.000$$

From the center of the arc to the start point A:

$$I = 0.000$$
$$J = 3.000$$

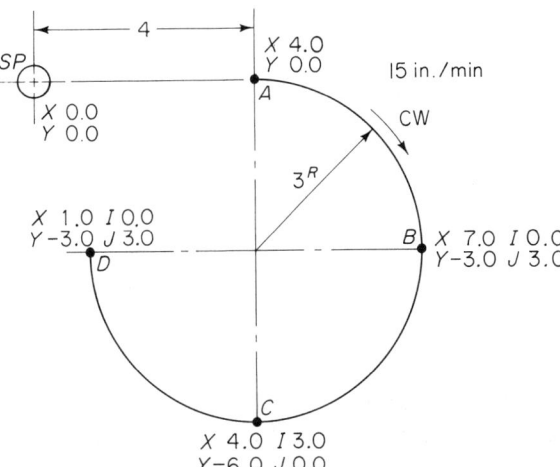

Figure 12.6

Sec. 12.2 Multiquadrant Circular Interpolation

At C:
$$X = 7.000 - 3.000 = 4.000$$
$$Y = -3.000 - 3.000 = -6.000$$

From the center of the arc to the start point B:
$$I = 3.000$$
$$J = 0.000$$

At D:
$$X = 4.000 - 3.000 = 1.000$$
$$Y = -6.000 + 3.000 = -3.000$$

From the center of the arc to the start point C:
$$I = 0.000$$
$$J = 3.000$$

(b) The program (Fig. 12.6) in the absolute mode is shown in Table 12.5(a).

(c) The program in the incremental mode is shown in Table 12.5(b). The movements are from point to point. Multicircular interpolation may also be accomplished using a G75 code.

TABLE 12.5(a)

N Address	G Code	X Axis	Y Axis	I Arc	J Arc	Feed (in./min)
N10	G90G01	4.0	0.0			150.0
N20	G74G02	7.0	-3.0	0.0	3.0	
N30		4.0	-6.0	3.0	0.0	
N40		1.0	-3.0	0.0	3.0	
N50	G00	0.0	0.0			

G90 indicates the absolute mode; G01 the feed mode; G02 clockwise rotation; G00 rapid; I and J unsigned arcs.

TABLE 12.5(b)

N Address	G Code	X Axis	Y Axis	I Arc	J Arc	Feed (in./min)
N10	G91G01	4.0	0.0			150.0
N20	G74G02	3.0	-3.0	0.0	3.0	
N30		-3.0	-3.0	3.0	0.0	
N40		-3.0	3.0	0.0	3.0	
N50	G00	-1.0	3.0			

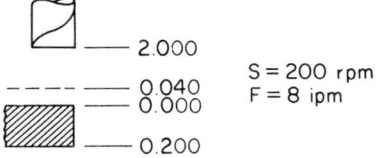
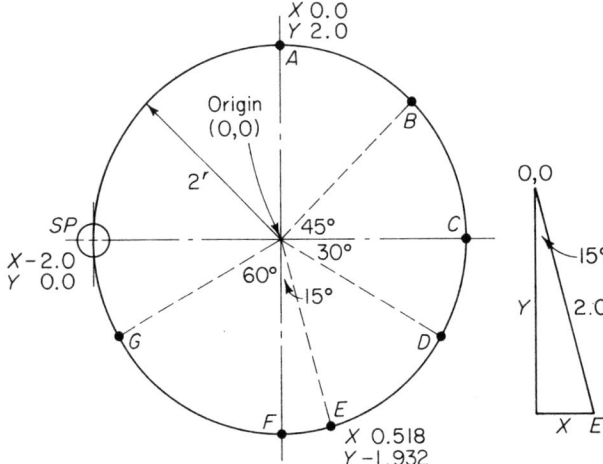

Figure 12.7

Example 6

Program Fig. 12.7 from *A* to *E* in the absolute mode using a G75 code. Mill to a depth of 0.200.

Solution (a) The coordinates at point *E* are

$$\cos 15° = \frac{Y}{2}$$

$$Y = 2 \cos 15° = 1.932$$

$$X = 2 \sin 15° = 0.518$$

The program (Fig. 12.7) is shown in Table 12.6.

Example 7

Fig. 12.8(a) is to be programmed in the G74 absolute mode. The feed is 15 in./min. and the spindle speed is 200 rev/min.

  (a) Do the analysis at each junction.
  (b) Redraw the part and insert all the X, Y, I, J codes at all junction points.
  (c) Program the part.

TABLE 12.6

N Address	G Code	X Axis	Y Axis	Z Axis	I Arc	J Arc	Spindle Speed (rev/min)	Feed (in./min)	M Function	Comment
N10	G00/G90	-2.0	0.0				2000		M6	At SP
N20	G75									Absolute mode
N30	G01	0.0	2.0	0.04						At A
N40				-0.2				8.0		At A
N50	G02	0.518	-1.932		0.0	0.0	(origin)			To E
N60	G00	-2.0	0.0							SP
N70									M2	(end of program)

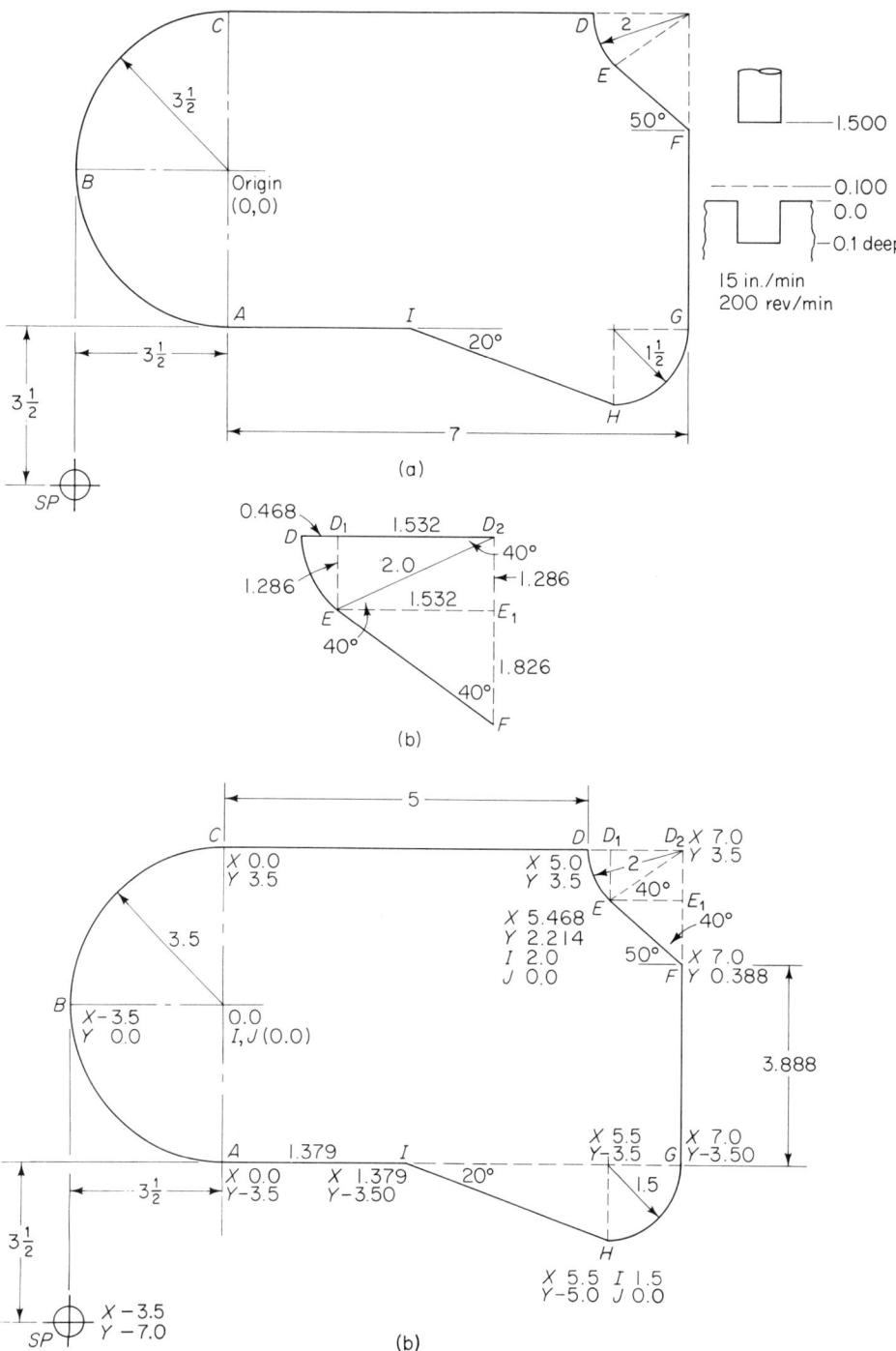

Figure 12.8

Solution (a) The analysis for Fig. 12.8(a) is:
Origin to SP:
$$X = -3.500$$
$$Y = -7.000$$
SP to A:
$$X = -3.500 + 3.500 = 0.000$$
$$Y = -7.000 + 3.500 = -3.500$$
A to C:
$$X = 0.000$$
$$Y = -3.500 + 7.000 = 3.500$$
C to D:
$$7.000 - 2.000 = 5.000$$
$$X = 0.000 + 5.000 = 0.500$$
$$Y = 3.500 + 0.000 = 3.500$$

In Fig. 12.8(b) and triangle ED_1D_2,

$$\sin 40° = \frac{ED_1}{2}$$

$$ED_1 = 2 \sin 40° = 1.286$$

$$\cos 40° = \frac{D_1D_2}{2}$$

$$D_1D_2 = 2 \cos 40° = 1.532 \text{ in.}$$

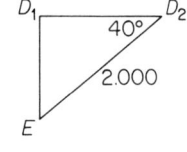

Therefore,
$$DD_1 = 2.000 - 1.532 = 0.468 \text{ in.}$$

D to E:
$$X = 5.000 + 0.468 = 5.468 \text{ in.} \qquad I = 2.000$$
$$Y = 3.500 - 1.286 = 2.214 \text{ in.} \qquad J = 0.000$$

E to F:
$$X = 5.468 + 1.532 = 7.000 \text{ in.}$$

$$\tan 40° = \frac{1.532}{E_1F}$$

$$E_1F = \frac{1.532}{\tan 40°} = 1.826 \text{ in.}$$

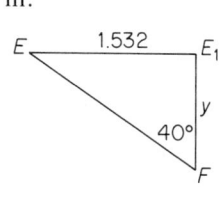

TABLE 12.7

N Address	G Code	X Axis	Y Axis	Z Axis	I Arc	J Arc	Feed (in./min)	Spindle Speed (rev/min)	M Function	Comment
%N10	G00 G90	-3.5	-7.0	1.5				2000	M06	At SP
N20	G74									At SP
N30							15.0			SP to A
N40	G01	0.0	-3.5	0.1			7.2			A to B
N50	G02	-3.5	0.0	-0.1	0.0	3.5				A to C
N60	G01	0.0	3.5							C to D
N70	G03	5.468	2.214		2.0	0.0				D to E
N80	G01	7.0	0.388							E to F
N90			-3.5							F to G
N100	G02	5.5	-5.0		1.5	0.0				G to H
N110	G01	1.379	-3.5							H to I
N120		0.0	-3.5							I to A
N130	G0			1.5						Up at A
N140		-3.5	-7.0							A to SP
N150									M02	(end of program)

Therefore,
$$Y = 2.214 - 1.862 = 0.388 \text{ in.}$$

F to G:
$X = 7.000$ in.
$FG = 8.500 - (1.286 + 1.826 + 1.500) = 3.888$ in.
$Y = 0.338 - 3.888 = -3.500$ in.

G to H:
$X = 7.000 - 1.500 = 5.500$ in. $I = 1.500$
$Y = -3.500 - 1.500 = -5.000$ in. $J = 0.000$

H to I:
$$\tan 20° = \frac{1.5}{I_1}$$
$$I_1 = \frac{1.5}{\tan 20°} = 4.121 \text{ in.}$$
$X = 5.500 - 4.121 = 1.379$ in.
$Y = -5.000 + 1.500 = -3.500$ in.

I to A:
$X = 1.379 - 7.000 - 1.500 - 4.121 = 0.000$
$Y = -3.500$ in.

A to $\oplus$:
$X = 0.000 - 3.500 = -3.500$ in.
$Y = -3.500 - 3.500 = -7.000$ in.

(b) The coded dimensions have been inserted into Fig. 12.8(b).
(c) The program for Fig. 12.8(b) is as shown in Table 12.7.

QUESTIONS AND PROBLEMS

12.1. Explain the I, J, K codes used in circular interpolation in this chapter.
12.2. Which codes are used for cutters that operate:
  (a) In a clockwise direction?
  (b) In a counterclockwise direction
12.3. Explain how I and J distances are determined.

12.4. Explain the X, Y, I, J values in Fig. 12.9.

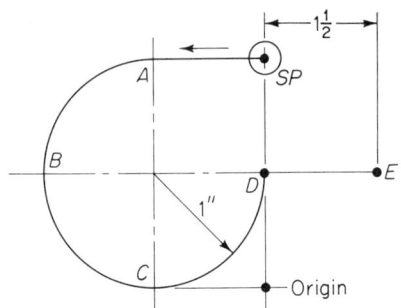

Figure 12.9

12.5. Develop the X, Y, I, J values in Fig. 12.10. Write the program in the absolute mode.

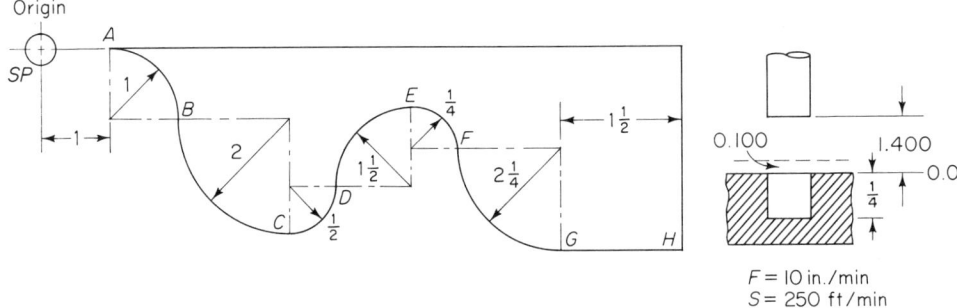

Figure 12.10

12.6. Program Fig. 12.11 in the absolute mode using multiquadrant circular interpolation.

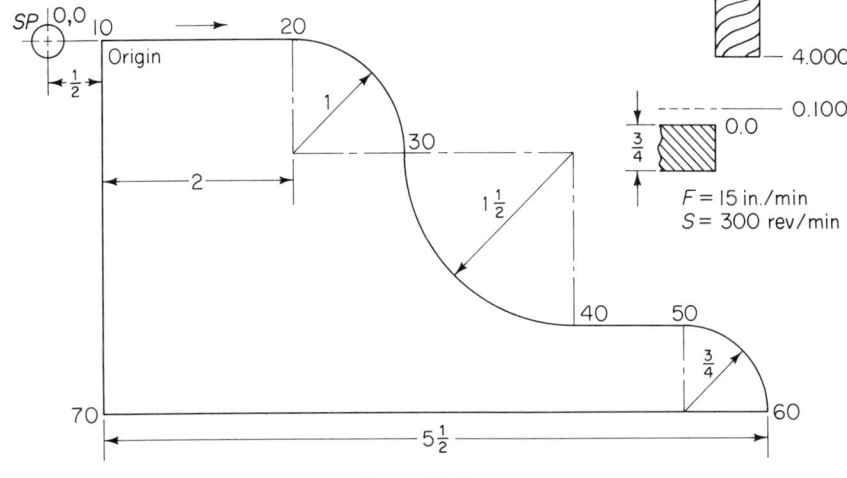

Figure 12.11

Chap. 12 Questions and Problems

12.7. Program Fig. 12.12 in the absolute mode using multiquadrant circular interpolation.

12.8. Program Fig. 8.12 in the absolute mode. Use multiquadrant circular interpolation.

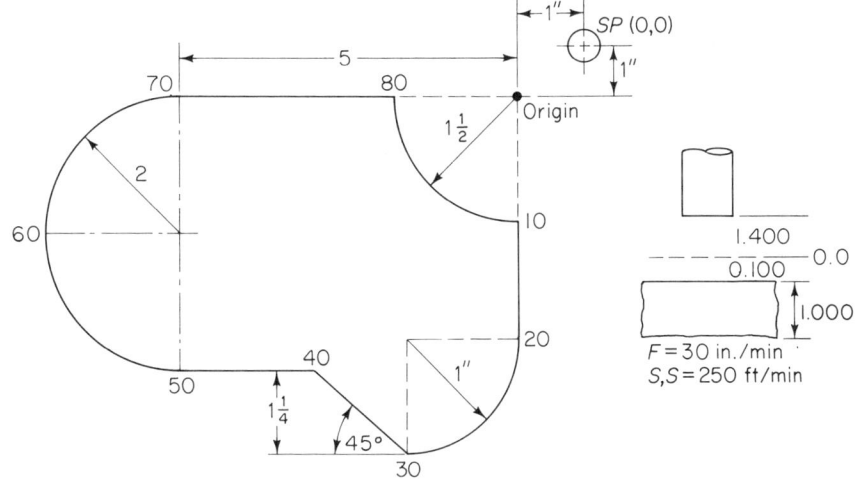

Figure 12.12

12.9. Program Fig. 8.13 in the absolute mode.
12.10. Program Fig. 8.14 in the absolute mode.
12.11. Repeat Prob. 12.9 for Fig. 8.15.
12.12. Repeat Prob. 12.9 for Fig. 8.16.
12.13. Use the G75 code and program all points in Fig. 12.7.
12.14. Repeat Prob. 12.13 for Fig. 15.6.

13

Radius Compensation: Milling Machine

13.1 RADIUS COMPENSATION

To this point all programming has been done as though the workpiece had been outlined with a pencil. However, if it is desired to use an end mill, the *radius* of the end mill must be considered when programming an operation.

13.2 NINETY-DEGREE MOVEMENTS

The simplest movement of an end mill is to cut either the outside of a rectangle [Fig. 13.1(a)] or the inside of a rectangle, (Fig. 13.2). At each corner the cutter is to generate a 90° move.

Example 1

The outside of Fig. 13.1(a) is to be machined with a $\frac{1}{2}$-in.-diameter end mill.

(a) Without correcting for the radius of the cutter, analyze the point-to-point movement of a $\frac{1}{2}$-in.-diameter cutter.

(b) Calculate the correction at each junction point and apply the correction to the X and Y dimensions.

(c) Write the program in the absolute mode.

Sec. 13.2 Ninety-degree Movements

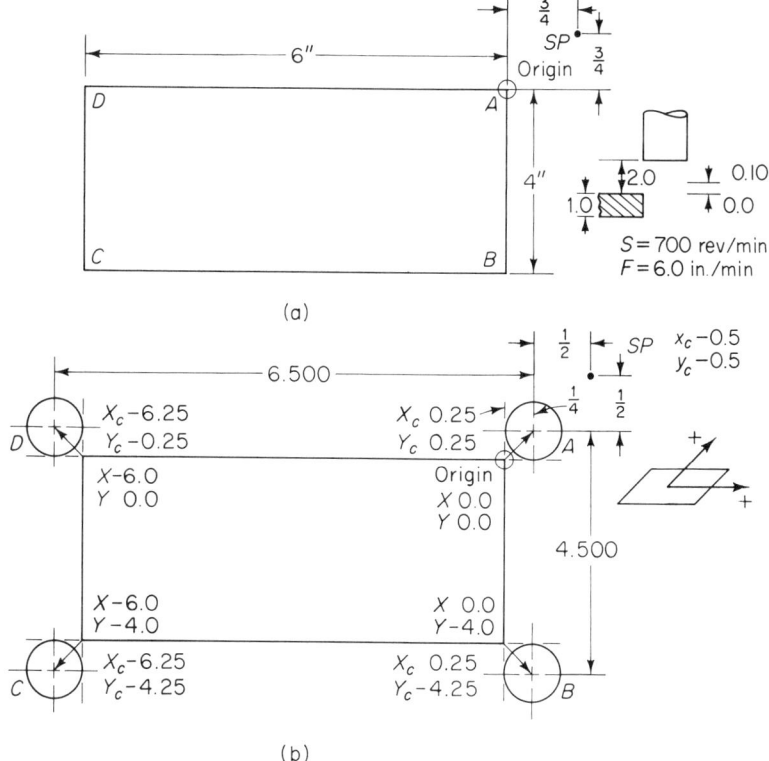

Figure 13.1

Solution (a) Point-to-point analysis. Draw the workpiece as shown in Fig. 13.1(b) and show the cutter compensation dimension. The first step is to determine the X, Y points *on the workpiece*. These should be inserted into Fig. 13.1(b).

At *A* (the origin):

$$X = 0.000$$
$$Y = 0.000$$

A to *B*:

$$X = 0.000$$
$$Y = 0.000 - 4.000 = -4.000$$

B to *C*:

$$X = 0.000 - 6.000 = -6.000$$
$$Y = -4.000 + 0.000 = -4.000$$

196   Radius Compensation: Milling Machine   Chap. 13

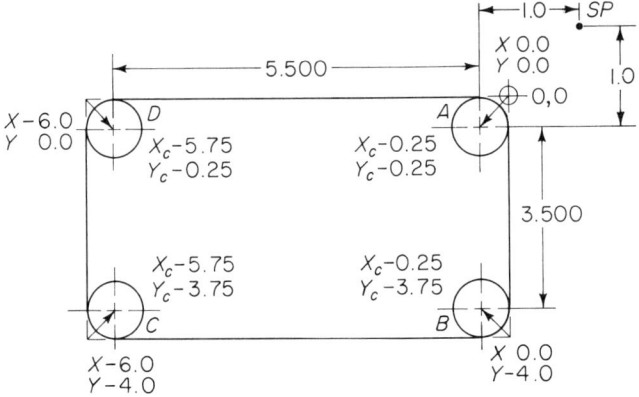

Figure 13.2

C to D:

$$X = -6.000 + 0.000 = -6.000$$
$$Y = -4.000 + 4.000 = 0.000$$

D to A:

$$X = -6.000 + 6.000 = 0.000$$
$$Y = 0.000$$

(b) The next step is to make the correction at each junction using $\frac{1}{4}$-in. radius of the cutter as shown in Fig. 13.1(b).

At A (corrected):

$$X_c = 0.000 + 0.250 = 0.250$$
$$Y_c = 0.000 + 0.250 = 0.250$$

A to B:

$$X_c = 0.000 + 0.250 = 0.250$$
$$Y_c = -4.000 - 0.250 = -4.250$$

B to C:

$$X_c = -6.000 - 0.250 = -6.250$$
$$Y_c = -4.000 - 0.250 = -4.250$$

Sec. 13.2 Ninety-degree Movements

C to D:

$$X_c = -6.000 - 0.250 = -6.250$$
$$Y_c = 0.000 + 0.250 = 0.250$$

D to A:

$$X_c = 0.000 + 0.250 = 0.250$$
$$Y_c = 0.000 + 0.250 = 0.250$$

(c) The program for Fig. 13.1(b) is shown in Table 13.1.

Example 2

Machine the *inside* of Fig. 13.2. The cutter diameter is $\frac{1}{2}$ in. Repeat Example 1.

Solution (a) The corner dimensions are the same as in Example 1.
(b) These dimensions are corrected at each junction point and inserted into the drawing, Fig. 13.2.

To A:

$$X_c = 0.000 - 0.250 = -0.250$$
$$Y_c = 0.000 - 0.250 = -0.250$$

A to B:

$$X_c = 0.000 - 0.250 = -0.250$$
$$Y_c = -4.000 + 0.250 = -3.750$$

B to C:

$$X_c = -6.000 + 0.250 = -5.750$$
$$Y_c = -4.000 + 0.250 = -3.750$$

C to D:

$$X_c = -6.000 + 0.250 = -5.750$$
$$Y_c = 0.000 - 0.250 = -0.250$$

D to A:

$$X_c = 0.000 - 0.250 = -0.250$$
$$Y_c = 0.000 - 0.250 = -0.250$$

(c) The program for Fig. 13.2 is shown in Table 13.2.

TABLE 13.1

N Address	G Code	X Axis	Y Axis	Z Axis	Feed (in./min)	Spindle Speed (rev/min)	T/H Tool	M Function	Comment
N10	G00G90	0.5	0.5	2.0				M06	Rapid at SP
N20						700	T01/H01	M03	Spindle on: CW
N30		0.25	0.25	0.1					SP to A
N40	G01			-0.5	6.0				To depth at A
N50			-4.25						A to B
N60		-6.25							B to C
N70			0.25						C to D
N80		0.25							D to A
N90	G00	0.75	0.75	2.0					Rapid to SP
N100								M02	

TABLE 13.2

N Address	G Code	X Axis	Y Axis	Z Axis	Feed (in./min)	Spindle Speed (rev/min)	T/H Tool	M Function	Comment
N10	G00G90	1.0	1.0	2.0			T01/H01	M06	Rapid at SP
N20						700		M03	Spindle on rapid
N30		-0.25	-0.25	0.1					SP to A
N40	G01			-0.5	6.0				To depth at A
N50			-3.75						A to B
N60		-5.75							B to C
N70			-0.25						C to D
N80		-0.25							D to A
N90	G00			2.0					Up at A
N100		1.0	1.0						Rapid to SP
N110								M02	(end of program)

200 Radius Compensation: Milling Machine   Chap. 13

13.3 ANGULAR COMPENSATION

There are four basic kinds of angular movements when milling that need tool radius compensation. They are shown in Fig. 13.3(a) to (c). Figure 13.3(d) shows the compensation necessary to cut a groove.

The cutter compensation will take place from A to A' and from B to B' in Fig. 13.4(a). The program correction will take place along the and axes as shown by the arrows. It should be noted that the distances AA' and BB' are *not* the radii of the cutter. However, since the angle 0° may be found, as shown in Fig. 13.4(b) and (c) and the radius of the cutter is known, it is possible to find the outer leg of the triangle.

Example 3

Compensate for the $\frac{1}{2}$-in.-diameter cutter shown in Fig. 13.4(b). What are the r and k compensation values when the center of the cutter is to be adjusted from A to A'?

Solution

1. The radius movement along the X axis is

$$r = 0.250$$

2. To find k, the correction factor along the X axis, it is noted that the exterior angle is given as 120°.

3. The angle θ° is

$$\theta° = 90° - \frac{120°}{2} = 30°$$

4. The correction k is

$$\tan 30° = \frac{k}{r} = \frac{k}{0.250}$$

$$k = 0.250 \tan 30°$$

$$= 0.144 \text{ in.}$$

The correction value for k and r will give the correction for k and r. They are

$$r = 0.250 \text{ in.} \quad \text{and} \quad k = 0.144 \text{ in.}$$

Example 4

In Fig. 13.4(c) calculate the correction k. Use a $\frac{1}{2}$-in.-diameter end mill.

Solution

1. The r movement in Fig. 13.4(c) is the radius of the cutter:

$$r = 0.250 \text{ in.}$$

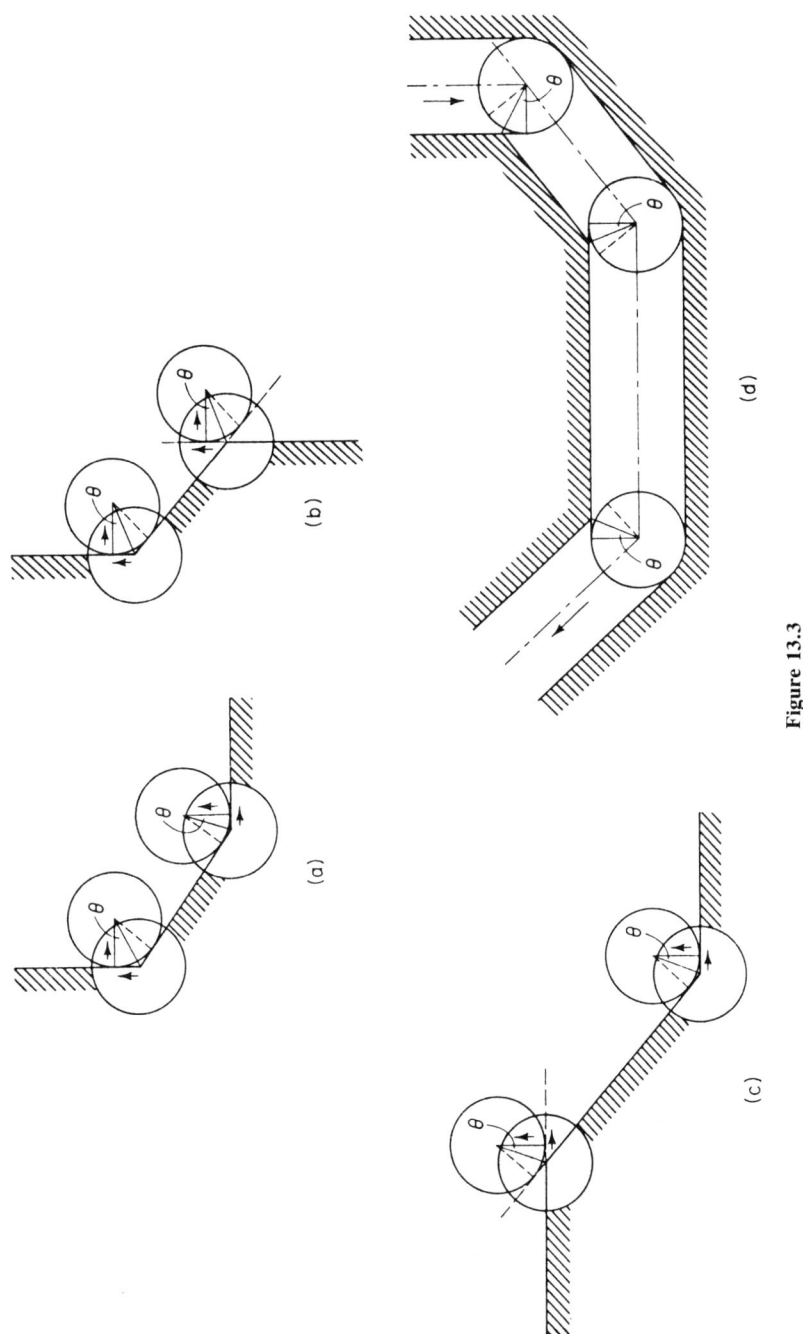

Figure 13.3

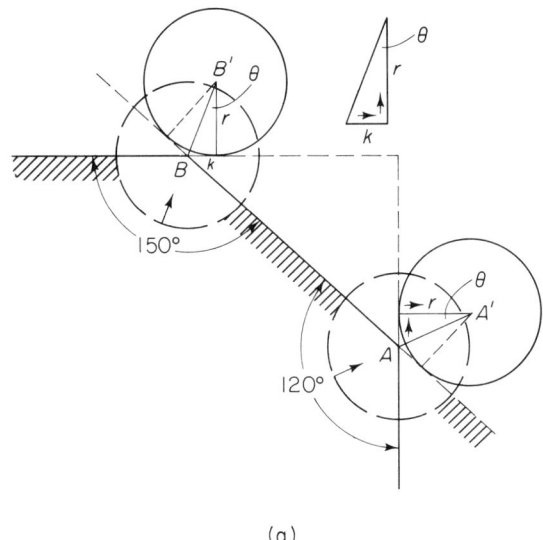

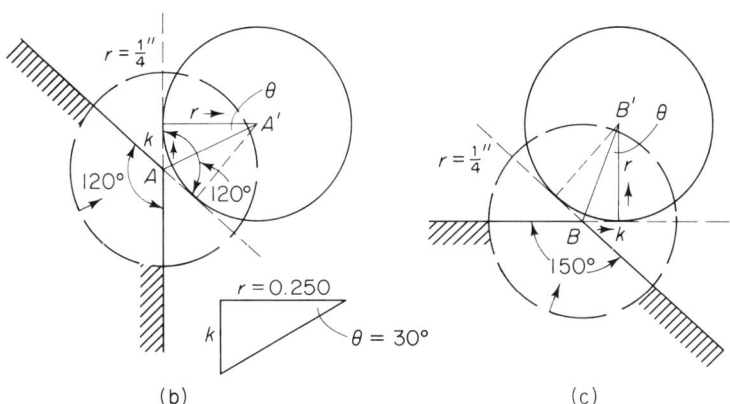

Figure 13.4

2. The angle $\theta°$ is

$$\theta° = 90° - \frac{150°}{2} = 15°$$

The k correction is

$$\tan 15° = \frac{k}{r} = \frac{k}{0.250}$$

$$k = 0.250 \tan 15°$$
$$= 0.067 \text{ in.}$$

Sec. 13.3 Angular Compensation

The correction values for k and r will give the values for k and r. They are

$$r = 0.250 \text{ in.} \quad \text{and} \quad k = 0.067 \text{ in.}$$

Example 5

The outside of Fig. 13.5 is to be machined with a 1-in.-diameter ($\frac{1}{2}$-in.-radius) end mill in the absolute mode.

(a) Analyze and redimension the drawing in the absolute mode when *no* compensation is applied.
(b) Apply radius compensation.
(c) Program the part.

Solution (a) Figure 13.5(a) is analyzed and redimensioned in Fig. 13.5(b) and prepared for absolute x and y coordinates prior to compensation. The analysis is as follows:

The components of *CD* are

$$\tan 30° = \frac{1.4}{x}$$

$$x = \frac{1.4}{\tan 30°} = 2.425$$

$$y = 1.400$$

The components of *FG* are

$$x = 2.000$$
$$y = 2.000$$

The components of *HI* are

$$\tan 20° = \frac{1.300}{x}$$

$$x = \frac{1.300}{\tan 20°} = 3.572$$

$$y = 1.300$$

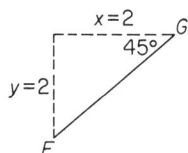

The following components may now be calculated:

$$CB = 12.000 - (2.500 + 2.425 + 5.000) = 2.075$$
$$FE = 3.800 - (2.000) = 1.800$$
$$I = 3.800 - (1.300) = 2.500$$
$$GH = 12.000 - (2.000 + 3.572) = 6.428$$

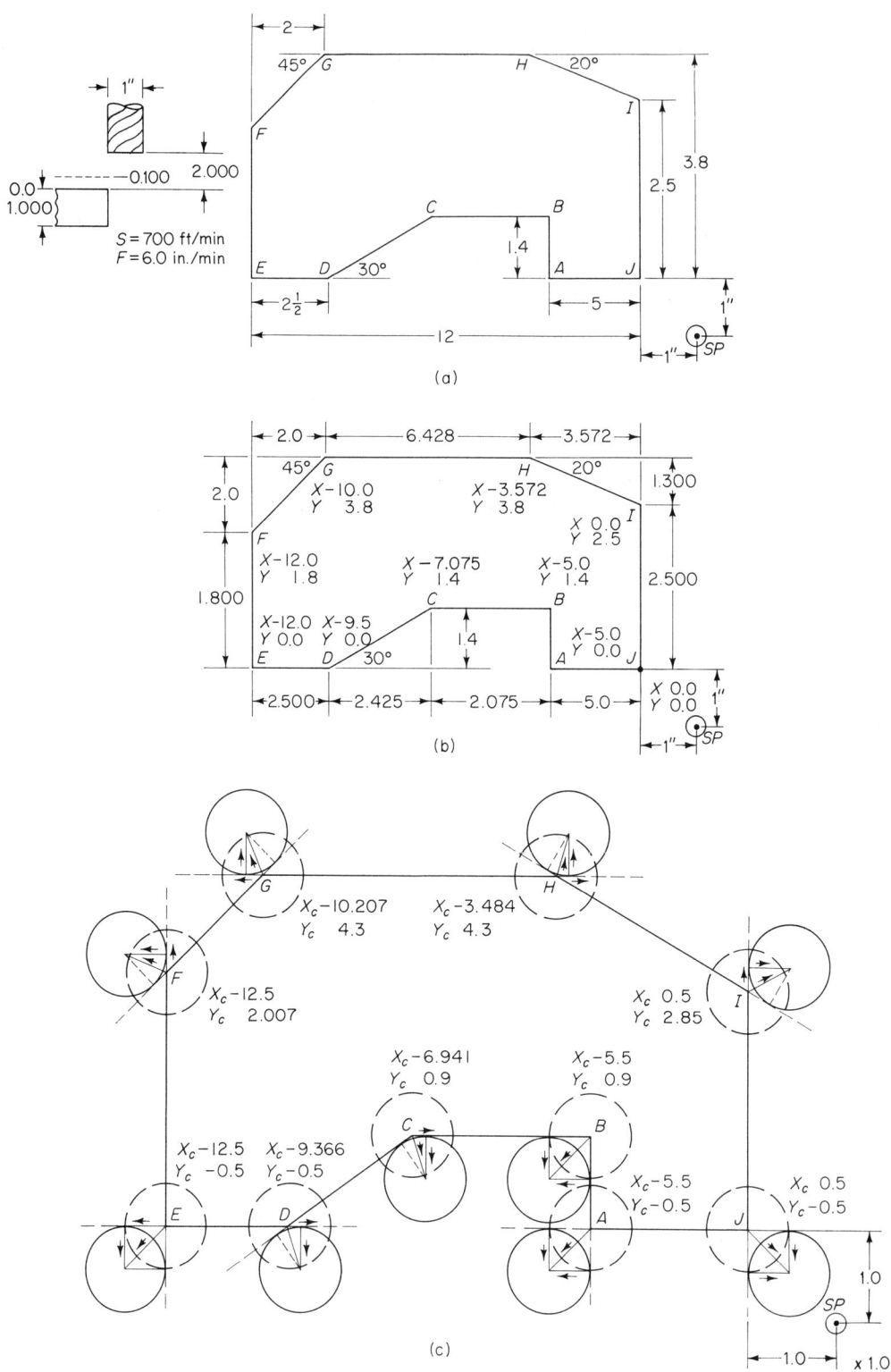

Figure 13.5

Sec. 13.3 Angular Compensation

The redimensioned drawing, if no compensation is applied, is as follows and they have been applied to Fig. 13.5(b).

J to A:

X = −5.000

Y = 0.000

B to C:

X = −5.000 − 2.075 = −7.075
Y = 1.400

D to E:

X = −9.500
Y = −12.000

F to G:

X = −12.000 + 2.000
= −10.000
Y = 1.800 + 2.000 = 3.800

H to I:

X = −3.572 − 3.572 = 0.000
Y = 3.800 − 1.300 = 2.500

A to B:

X = −5.000

Y = 0.000 + 1.400 = 1.400

C to D:

X = −7.075 − 2.425 = −9.500
Y = 1.400 − 1.400 = 0.000

E to F:

X = −2.000
Y = 0.000 + 1.800 = 1.800

G to H:

X = −10.000 + 6.428
= −3.572
Y = 3.800

I to J:

X = 0.000
Y = 2.500 − 2.500
= 0.000

(b) Apply radius cutter compensation to each junction. The cutter diameter is 1 in. The coordinates that follow have been applied to Fig. 13.5(c).

At J:

K = 0.500
X_c = X + K
= 0.000 + 0.500 = 0.500
Y_c = Y + K
= 0.000 − 0.500 = −0.500

A to B:

K = 0.500
X_c = −5.000 − 0.500
= −5.500
Y_c = 1.400 − 0.500
= 0.900

J to A:

K = 0.500
X_c = −5.000 − 0.500
= −5.500
Y_c = 0.000 − 0.500
= −0.500

B to C:

$\theta° = 90° - \dfrac{180 - 30}{2}$

= 15°

K = 0.500 tan 15° = 0.134
X_c = −7.075 + 0.134
= −6.941
Y_c = 1.400 − 0.500 = 0.900

TABLE 13.3

N Address	G Code	X Axis	Y Axis	Z Axis	Feed (in./min)	Spindle Speed (rev/min)	T/H Tool	M Function	Comment
N10	G00G90	1.0	-1.0	2.0			T01	M06	Rapid at SP
N20						700		M03	Spindle on: CW
N30		0.5	-0.5	0.1					SP to J
N40	G01			-0.5	6.0				AJ
N50		-5.5							J to A
N60			0.9						A to B
N70		-6.941							B to C
N80		-9.366	-0.5						C to D
N90		-12.5							D to E
N100			2.007						E to F
N110		-10.207	4.3						F to G
N120		-3.484							G to H
N130			2.85						H to I
N140		0.5	-0.5						I to J
N150	G00			2.0					At J
N160		1.0	-1.0						Rapid J to SP
N170								M02	(end of program)

Sec. 13.4 Circular Interpolation

C to *D*:

$$\theta° = 90° - \frac{180° - 30°}{2} = 15°$$
$$K = 0.500 \tan 15°$$
$$= 0.134$$
$$X_c = -9.500 + 0.134 = -9.366$$
$$Y_c = 0.000 - 0.500 = -0.500$$

D to *E*:

$$X_c = -12.000 - 0.500$$
$$= -12.500$$
$$Y_c = 0.000 - 0.500$$
$$= -0.500$$

E to *F*:

$$\theta° = 90° - \frac{180° - 45°}{2} = 22.5°$$
$$K = 0.500 \tan 22.5°$$
$$= 0.207$$
$$X_c = -12.000 - 0.500$$
$$= -12.500$$
$$Y_c = 1.800 + 0.207 = 2.007$$

F to *G*:

$$K = 0.207$$
$$X_c = -10.000 - 0.207$$
$$= -10.207$$
$$Y_c = 3.800 + 0.500$$
$$= 4.300$$

G to *H*:

$$\theta° = 90° - \frac{180° - 20°}{2} = 10°$$
$$K = 0.500 \tan 10° = 0.088$$
$$X_c = -3.572 + 0.088$$
$$= -3.484$$
$$Y_c = 3.800 + 0.500$$
$$= 4.300$$

H to *I*:

$$\theta° = 90° - \frac{180° - 70°}{2} = 35°$$
$$K = 0.500 \tan 35° = 0.350$$
$$X_c = 0.000 + 0.500 = 0.500$$
$$Y_c = 2.500 + 0.350 = 2.850$$

I to *J*:

$$X_c = 0.000 - 0.500 = 0.500$$
$$Y_c = 0.000 + 0.500 = -0.500$$

(c) The program for Fig. 13.5(c) is shown in Table 13.3.

13.4 CIRCULAR INTERPOLATION

This radius compensation for the milling machine may be accomplished by dimensioning the center of the cutter as shown in Fig. 13.6. The compensations are in the directions of the arrows. Thus in Fig. 13.6(a) and (b) the compensation from A to B is accomplished by adding, or subtracting, an amount equal to one half the tool diameter in the X and Y directions. Because the starting and ending points are changed, the I and J are changed in the same way.

In Fig. 13.6(c) the correction from A to B is one radius, in *either*

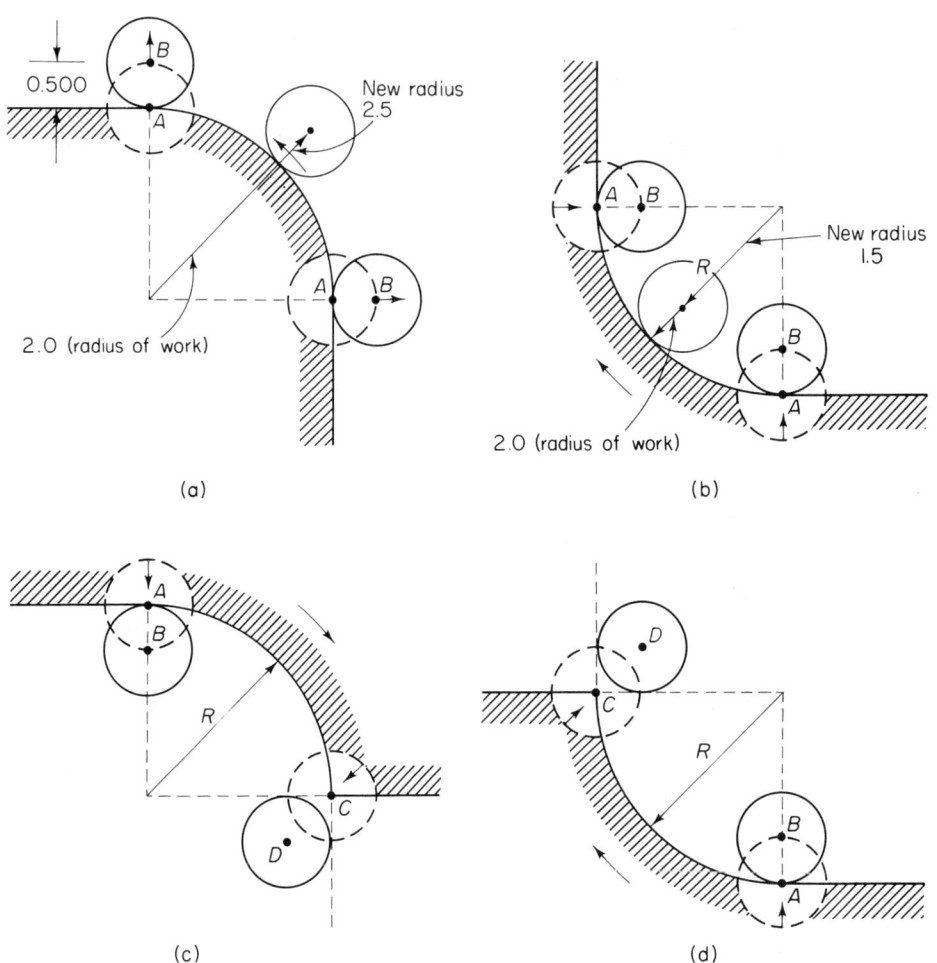

Figure 13.6

the X or Y direction. From C to D the correction is taken in *both* the X and Y directions.

Example 6

Using a ½ inch diameter cutter, program the junctions in Fig. 13.7(a) in the absolute mode. Use a spindle speed of 200 rpm, a depth of cut of 0.350 in. and a feed of 8 inches per minute.

    (a) Analyze and redimension the drawing and insert the dimensions without considering radius compensation.

    (b) Calculate the compensations at each junction point. Make

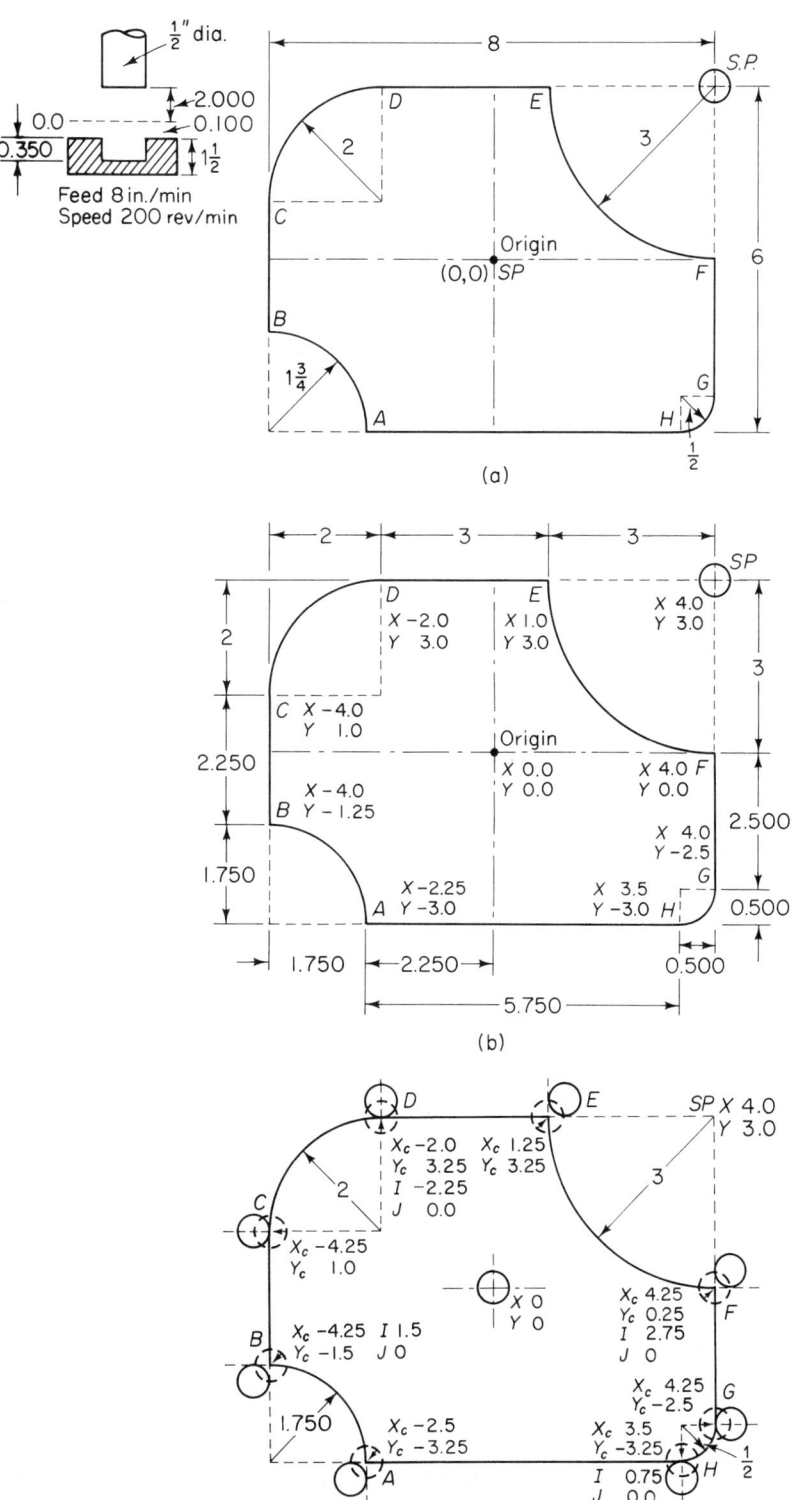

Figure 13.7

210   Radius Compensation: Milling Machine   Chap. 13

another drawing showing all compensations at each junction point.

(c) Program Fig. 13.7(a) with radius compensation.

Solution (a) The analysis is shown below. The redimensioned drawing is shown in Fig. 13.7(b). The noncompensated X, Y codes are also shown. They were arrived at in the following manner.

Origin to A:

$X = -2.250$

$Y = -2.500 - 0.500 = -3.000$

A to B:

$X = -2.250 - 1.750 = -4.000$

$Y = -3.000 + 1.750 = -1.250$

B to C:

$X = -4.000$

$Y = -1.250 + 2.250 = 1.000$

C to D:

$X = -4.000 + 2.000 = -2.000$

$Y = 1.000 + 2.000 = 3.000$

D to E:

$X = -2.000 + 3.000 = 1.000$

$Y = 3.000$

E to F:

$X = 1.000 + 3.000 = 4.000$

$Y = 3.000 - 3.000 = 0.000$

F to G:

$X = 4.000$

$Y = 0.000 - 2.500 = -2.500$

G to H:

$X = 4.000 - 0.500 = 3.500$

$Y = -2.500 - 0.500 = -3.000$

H to A:

$X = 3.500 - 5.750 = -2.250$

$Y = -3.000$

A to origin:

$X = -2.250 + 2.250 = 0.000$

$Y = -3.000 + 3.000 = 0.000$

(b) The compensated dimensions are shown below. They have been inserted into Fig. 13.7(c) and shown in Table 13.4.

SP to A

$X_c = -2.250 - 0.250 = -2.500$

$Y_c = -3.000 - 0.250 = -3.250$

A to B

$X_c = -4.000 - 0.250 = -4.250$

$Y_c = -1.250 - 0.250 = -1.500$

B to C

$X_c = -4.000 - 0.250 = -4.250$

$Y_c = -1.000$

C to D

$X_c = -2.000$

$Y_c = 3.000 + 0.250 = 3.250$

D to E

$X_c = 1.000 + 0.250 = 1.250$

$Y_c = 3.000 + 0.250 = 3.250$

E to F

$X_c = 4.000 + 0.250 = 4.250$

$Y_c = 0.000 + 0.250 = 0.250$

TABLE 13.4

N Address	G Code	X Axis	Y Axis	Z Axis	I Arc	J Arc	F (in/min)	S rpm	T/H Tool	M Function	Comment
N10	G00G90	4.0	3.0	2.0					T01/H01	M06	Rapid at SP
N20								200		M03	CW
N30		-2.5	-3.25	0.1							SP to A
N40	G01			-0.35			8.0				At A
N50	G03	-4.25	-1.5		1.5	0.0					A to B
N60	G01		1.0								B to C
N70	G02	-2.0	3.25		2.25	0.0					C to D
N80	G01	1.25									D to E
N90	G03	4.25	0.25		2.75	0.0					E to F
N100	G01		-2.5								F to G
N110	G02	3.5	-3.25		0.75	0.0					G to H
N120	G01	-2.5									H to A
N130	G00			2.0							Rapid up
N140		4.0	3.0								A to SP
N150										M02	EOP

F to G

$X_c = 4.000 + 0.250 = 4.250$

$Y_c = -2.500$

G to H

$X_c = 3.500$

$Y_c = -3.00 - 0.250 = -3.250$

H to A

$X_c = -2.250 - 0.250 = -2.500$

$Y_c = -3.000 - 0.250 = -3.250$

The program for Fig. 13.7(a) is shown in Table 13.4.

QUESTIONS AND PROBLEMS

13.1. Define radius compensation.

13.2. Define angular compensation.

13.3. Describe the process called "circular interpolation."

13.4. Use a 1-inch end mill and cut the outside of Fig. 7.7(a). Use the absolute mode.

13.5. Assume a $\frac{1}{2}$ inch diameter end mill to machine the *inside* of Fig. 13.5(a). Calculate the correction and the X, Y dimension at each junction point.

13.6. Draw Fig. 13.6(a) through (d) and explain the compensation at each junction point.

13.7. Program Fig. 13.8, using a 1-inch diameter end mill.

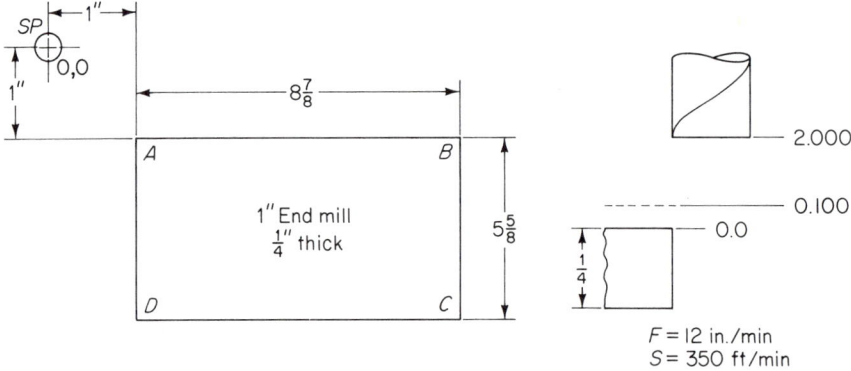

Figure 13.8

13.8. Program Fig. 13.9 using a 1-inch diameter end mill. The part is $\frac{1}{2}$ inch thick. Apply all necessary compensation.

Chap. 13 Questions and Problems

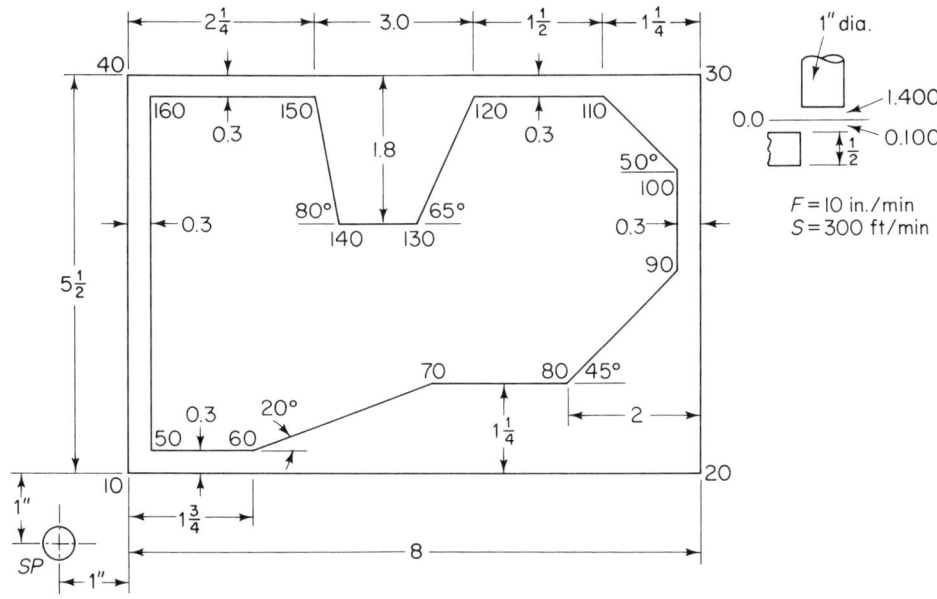

Figure 13.9

13.9. Program Fig. 13.10 using a $\frac{3}{4}$ inch end mill. The part is $\frac{1}{2}$ inch thick. The spindle speed is 250 rpm and the feed is 0.040 inches per minute.

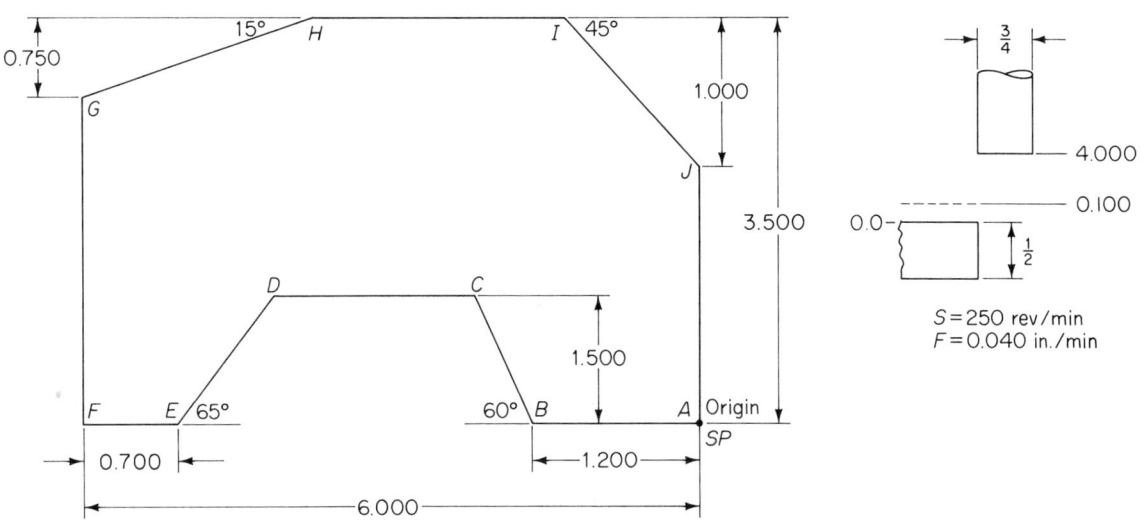

Figure 13.10

13.10. Program Fig. 13.11. The feed is 0.050 in/min. and the spindle speed is 300 ft/min.

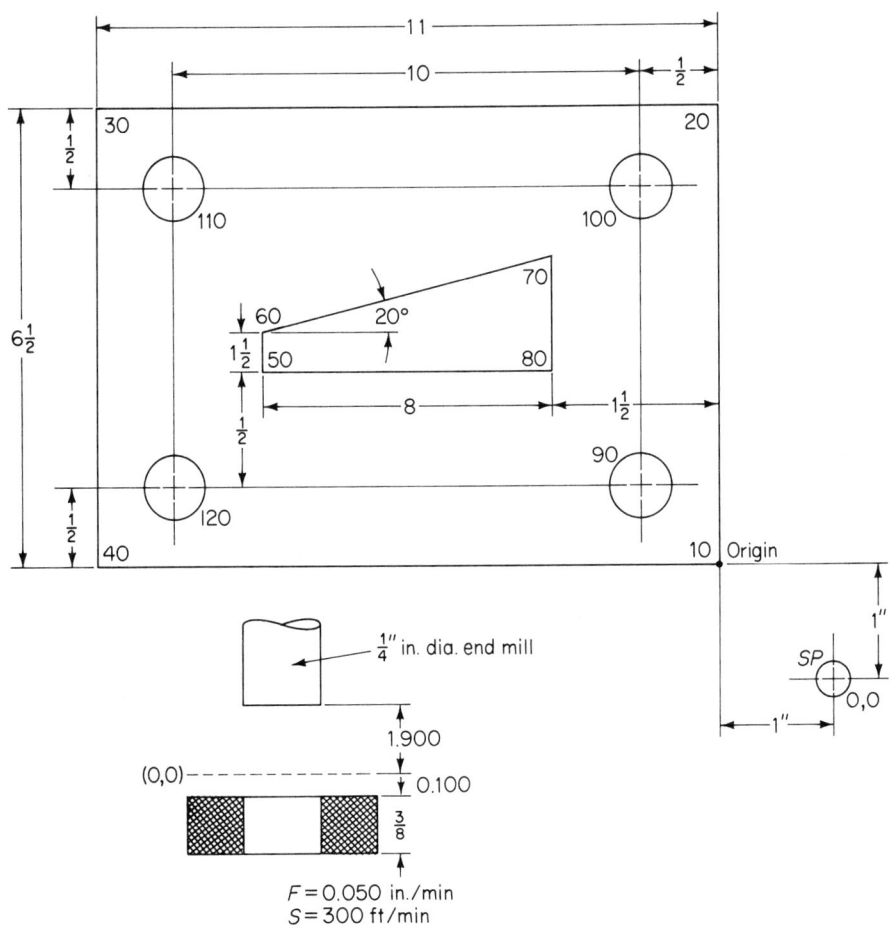

Figure 13.11

13.11. Using a ½-inch diameter end mill, program the junction points in Fig. 13.7(a). Assume that you are milling the inside of the part. Program the part if the workpiece is ½ in. thick.

13.12. Program Fig. 8.15 using a 1-inch diameter end mill. The part is ¾ in. thick.

13.13. Assume that the inside of Fig. 8.15 is to be programmed using a 1-in. diameter end mill. Program the part.

13.14. Using a ½ in. diameter end mill, program Fig. 12.10(a). The workpiece is ⅜ in. thick.

13.15. Repeat Problem 13.14 for the inside of Fig. 12.10(a).

14

Radius–Angle Combinations

14.1 RADIUS–ANGLE COMPENSATION

Other compensations that need consideration are radius–angle, radius–angle–radius, and so on. The corrections at points A, B, and C are shown in Fig. 14.1(a) and enlarged in Fig. 14.1(b) through (d).

1. Thus, in Fig. 14.1(b) at A, the correction is (use lowercase x and y)

$$x_c = 0$$

$$y_c = r$$

2. In Fig. 14.1(c) at B, the correction is

$$x_c = r \cos 30°$$

$$y_c = r \sin 30°$$

3. Figure 14.1(d) at C, the correction is

$$x_c = r \tan 30°$$

$$y_c = r$$

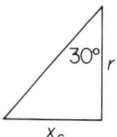

Example 1

Figure 14.2(a) is to be programmed for a $\frac{1}{2}$-in.-diameter end mill.

(a) Analyze Fig. 14.2(a) and redimension the drawing.

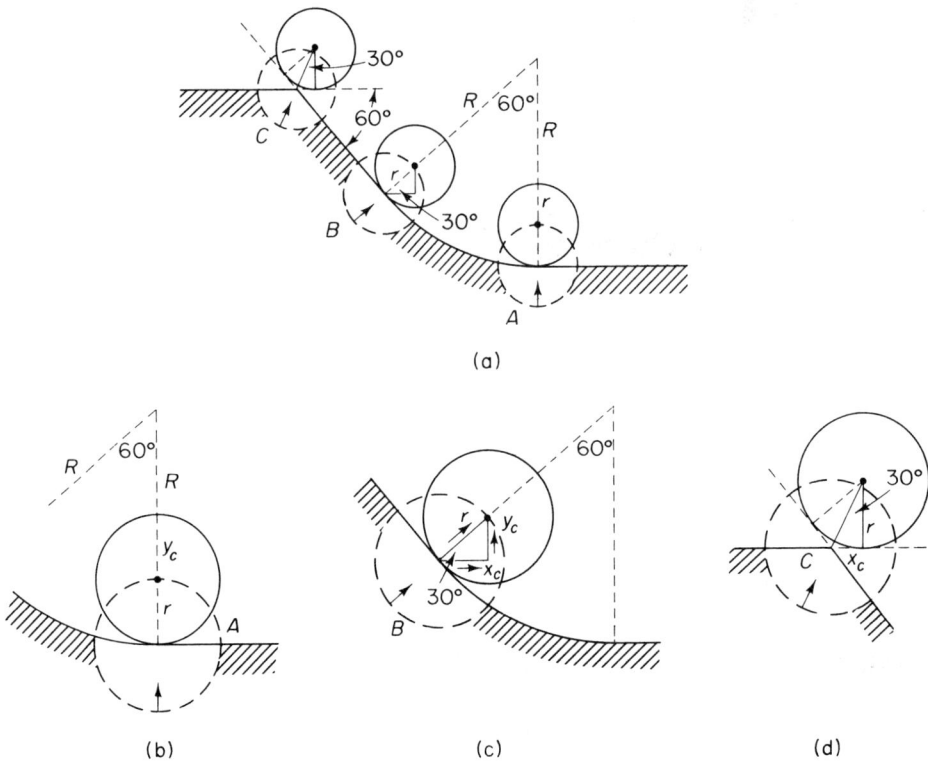

Figure 14.1

(b) Calculate and insert the absolute dimensions into the drawing *without compensation*.

(c) Calculate the compensated dimensions and insert them into a drawing.

(d) Write the program using a feed of 30 in./min and a spindle speed of 350 rev/min. Use G74 mode.

Solution (a) The analysis for Fig. 14.2(a) is: In Fig. 14.2(b) the radius generated by the arc CD is 2.500 in. The triangle $RC'D$ is a 30°–60°–90° triangle. Thus in triangle $RC'D$,

$$x = 2.500 \cos 30° = 2.165$$
$$y = 2.500 \sin 30° = 1.250$$

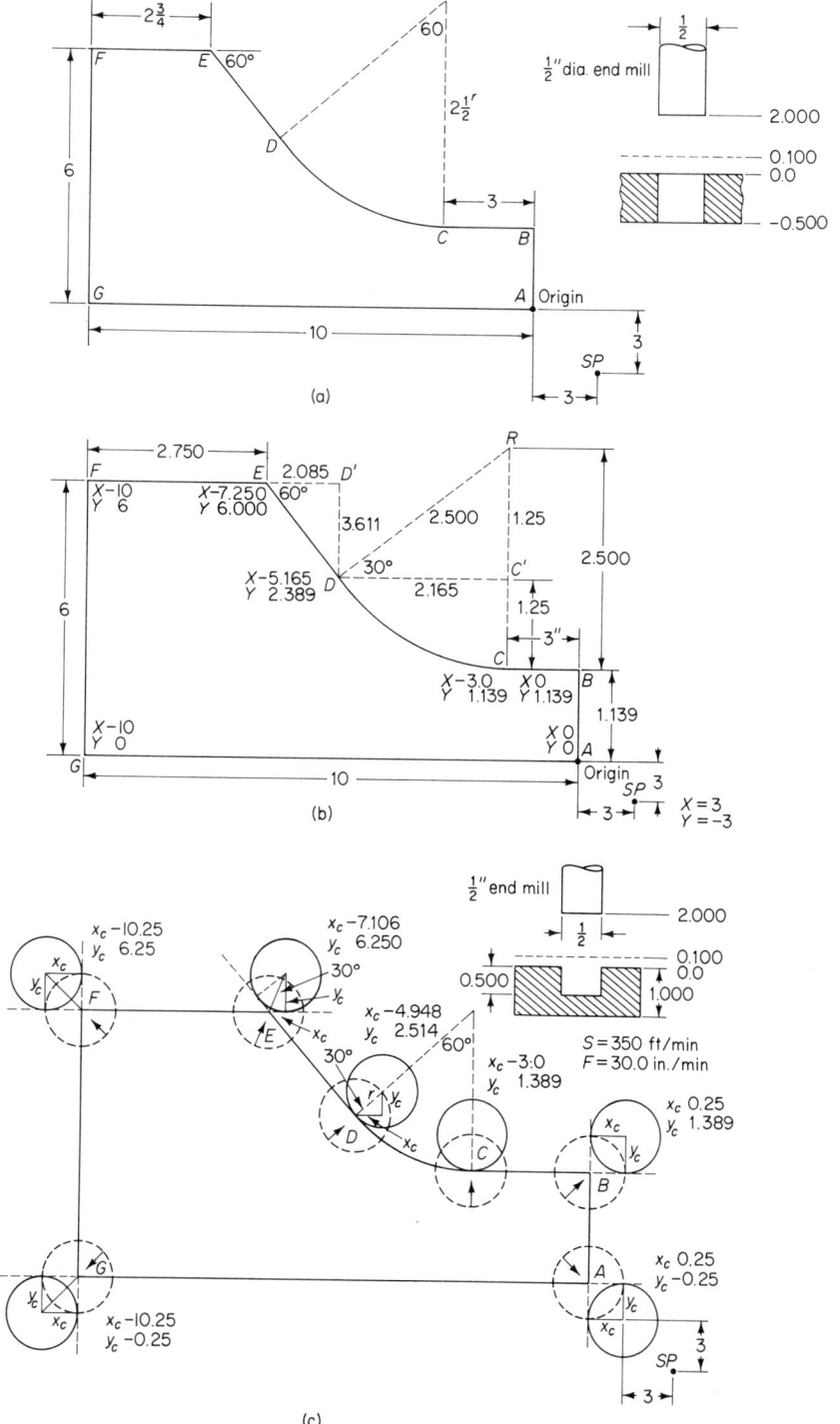

Figure 14.2

In triangle $DD'E$,

$$ED' = 10.000 - (2.750 + 2.165 + 3.000) = 2.085$$

and

$$DD' = 2.085 \tan 60° = 3.611$$

(b) The absolute XY dimensions are inserted into Fig. 14.2(b) as follows:

At SP:
$$X = 3.000$$
$$Y = -3.000$$

SP to A:
$$X = 0.000$$
$$Y = 0.000$$

A to B:
$$X = 0.000$$
$$Y = 0.000 + 1.139 = 1.139$$

B to C:
$$X = 0.000 - 3.000 = -3.000$$
$$Y = 1.139$$

C to D:
$$X = -3.000 - 2.165 = -5.165$$
$$Y = 1.139 + 1.250 = 2.389$$

D to E:
$$X = -5.165 - 2.085 = -7.250$$
$$Y = 2.389 + 3.611 = 6.000$$

E to F:
$$X = -7.250 - 2.750 = -10.000$$
$$Y = 6.000$$

F to G:
$$X = -10.000$$
$$Y = 6.000 - 6.000 = 0.000$$

Sec. 14.1   Radius–Angle Compensation   219

G to A:
$$X = -10.000 + 10.000 = 0.000$$
$$Y = 0.000$$

At SP:
$$X = 3.000$$
$$Y = -3.000$$

(c) The compensated dimensions are [see Fig. 14.2(c)] as follows:

At SP:
 $X = 3.000$
 $Y = -3.000$

SP to A:
 $X = 3.000 - 2.750 = 0.250$
 $Y = -3.000 + 2.750 = -0.250$

A to B:
 $X = 0.250$
 $Y = 1.139 + 0.250 = 1.389$

B to C:
 $X = 0.000 - 3.000 = -3.000$
 $Y = 1.389$

C to D:
 $X = -5.165 + 0.250 \cos 30° = -4.948 \qquad I = 0.000$
 $Y = 2.389 + 0.250 \sin 30° = 2.514 \qquad J = 2.250$

D to E:
 $X = -7.250 + 0.250 \tan 30° = -7.106$
 $Y = 6.000 + 0.250 = 6.250$

E to F:
 $X = -10.000 - 0.250 = -10.250$
 $Y = 6.250$

F to G:
 $X = -10.250$
 $Y = 0.000 - 0.250 = -0.250$

TABLE 14.1

N Address	G Code	X Axis	Y Axis	Z Axis	I Arc	J Arc	Spindle Speed (rev/min)	Feed (in./min)	T/H Tool	M Function	Comment
N10	G00G90	3.0	-3.0	2.0					T01/H01	M06	Load end at SP
N20							350			M03	Spindle on
N30		0.25	-0.25	0.1							SP to A
N40				-0.5							Full depth
N50	G01	0.25	1.389					30.0			A to B
N60		-3.0									B to C
N70	G74G02	-4.948	2.514		0.0	2.25					C to D
N80	G01	-7.106	6.25								D to E
N90		-10.25									E to F
N100			-0.25								F to G
N110		0.25									G to A
N120	G00			2.0							Rapid at A
N130		3.0	-3.0								A to SP
N140										M02	(end of program)

Sec. 14.2 Radius–Angle–Radius Compensation

G to A:

X = 0.000 + 0.250 = 0.250

Y = −0.250

A to SP:

X = 3.000

Y = −3.000

(d) The program for Fig. 14.2(a) is shown in Table 14.1.

14.2 RADIUS–ANGLE–RADIUS COMPENSATION

These types of calculations are an extension of the radius–angle programs. The corrections are added, or subtracted, from the dimensions at the junctions. The student should make two drawings of the part: one for insertion of the corrections at each junction and one for the final X and Y values at each junction. It is a good practice to use lowercase letters x and y for the corrections and capital letters X and Y with a sub-c (X_c and Y_c) for the corrected dimensions.

Example 2

(a) Calculate the X and Y dimensions for Fig. 14.3(a) at each junction. Insert these into a new drawing of the part.
(b) Calculate the x and y corrections at each junction.
(c) Make a new drawing of the part. Make the corrections at each junction and insert them into the drawing.
(d) Program the part.

Solution (a) The dimensions at each junction point are [see Fig. 14.3(b)]:

At SP:

X = 3.000

Y = −3.000

SP to A:

X = 0.000

Y = 0.000

A to B:

X = 0.000

Y = 3.000

222   Radius–Angle Combinations   Chap. 14

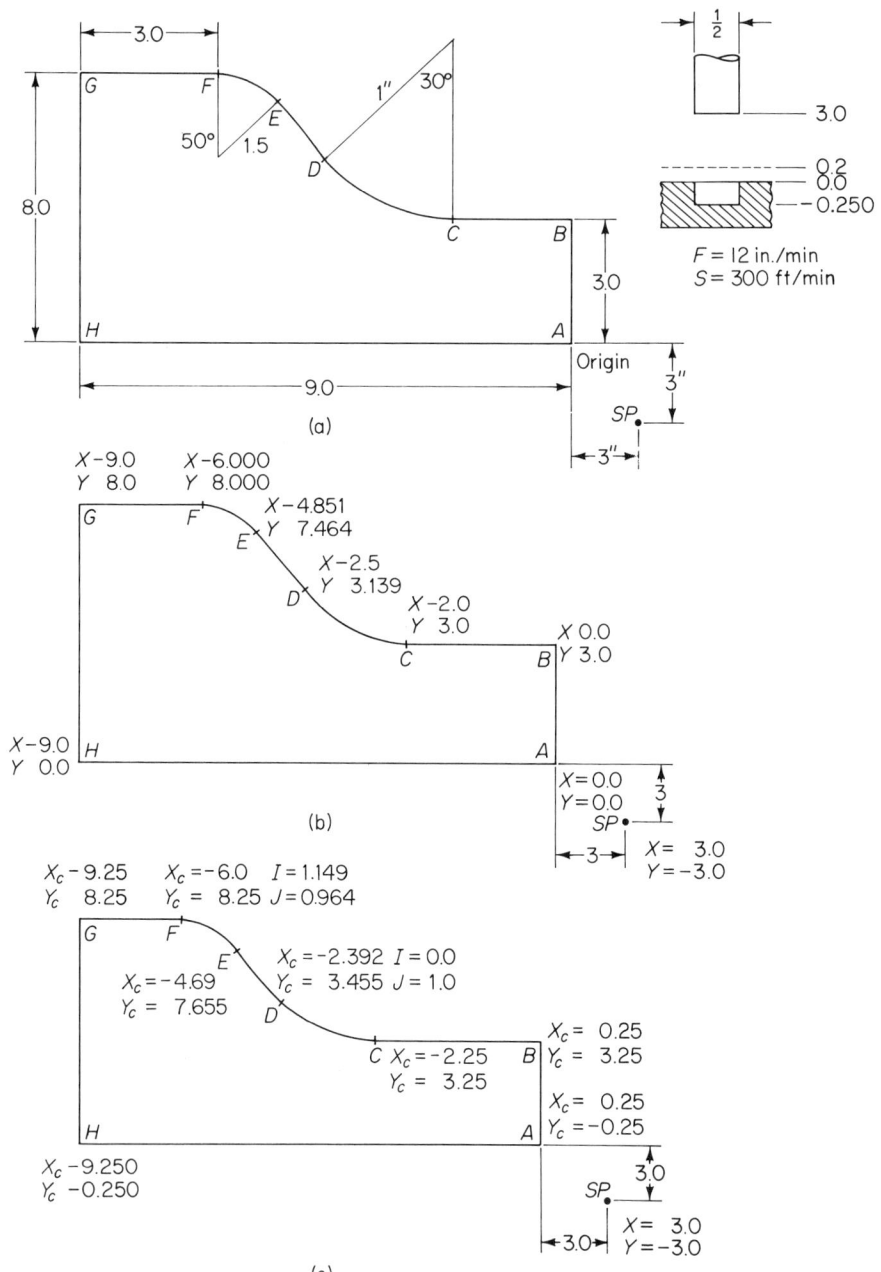

Figure 14.3

Sec. 14.2 Radius–Angle–Radius Compensation

B to C:

$X = -2.000$

$Y = 3.000$

C to D:

$X = -2.000 - 0.500 = -2.500 \quad I = 0.000$

$Y = 3.000 + 0.139 = 3.139 \quad J = -1.000$

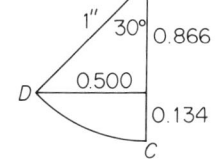

D to E:

$x = 9.000 - (2.000 + 0.500 + 1.149 + 3.000) = 2.351$

$y = 8.000 - (3.000 + 0.139 + 0.536) = 4.325$

Then

$X = -2.500 - 2.351 = -4.851$

$Y = 3.139 + 4.325 = 7.464$

E to F:

$X = -4.851 - 1.149 = -6.000 \quad I = 1.149$

$Y = 7.464 + 0.536 = 8.000 \quad J = 0.536$

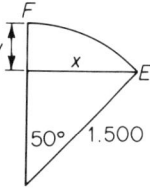

F to G:

$X = -6.000 - 3.000 = -9.000$

$Y = 8.000$

G to H:

$X = -9.000$

$Y = 8.000 - 8.000 = 0.000$

H to A:

$X = -9.000 + 9.000 = 0.000$

$Y = 0.000$

A to SP:

$X = 3.000$

$Y = -3.000$

These dimensions have been inserted into Fig. 14.3(b).
(b) The corrections at each junction are
At A:

$$x = 0.250$$
$$y = -0.250$$

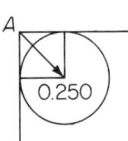

At B:

$$x = 0.250$$
$$y = 0.250$$

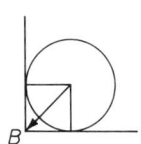

At C:

$$x = -0.250$$
$$y = 0.250$$

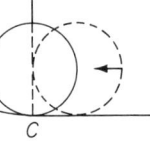

At D:

$$x = 0.250 \cos 60° = 0.108$$
$$y = 0.250 \sin 60° = 0.216$$

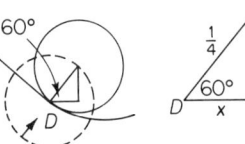

At E:

$$x = 0.250 \cos 50° = 0.161$$
$$y = 0.250 \sin 50° = 0.191$$

At F:

$$x = \text{no correction}$$
$$y = +0.250$$

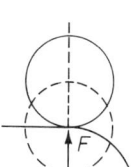

At G:

$$x = -0.250$$
$$y = 0.250$$

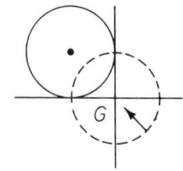

Sec. 14.2   Radius–Angle–Radius Compensation

At *H*:

$$x = -0.250$$
$$y = -0.250$$

(c) The corrected dimensions are as follows:

Move	X_c			Y_c		
SP to A			3.000			-3.000
At A	0.000	+ 0.250 =	0.250	0.000	- 0.250 =	-0.250
A to B	0.000	+ 0.250 =	0.250	3.000	+ 0.250 =	3.250
B to C	-2.000	- 0.250 =	-2.250	3.000	+ 0.250 =	3.250
C to D	-2.500	+ 0.108 =	-2.392	3.139	+ 0.216 =	3.455
D to E	-4.851	+ 0.101 =	-4.690	7.464	+ 0.191 =	7.655
E to F	-6.000		= -6.000	8.000	+ 0.250 =	8.250
F to G	-9.000	- 0.250 =	-9.250	8.000	+ 0.250 =	8.250
G to H	-9.000	- 0.250 =	-9.250	0.000	- 0.250 =	-0.250
H to A	0.000	+ 0.250 =	0.250	0.000	- 0.250 =	-0.250
SP to A	3.000		= 3.000	-3.000		= -3.000

There are two I and J dimensions that must be considered.
The arc *C* to *D*:

$$\sin 30° = \frac{X}{1.000}$$

$$X = 1.000 \sin 30° = 0.500 \quad I = 0.000$$
$$J = 1.000$$

The arc *E* to *F*:

$$\sin 50° = \frac{x}{1.500}$$

$$X = 1.500 \sin 50° = 1.149 \quad I = 1.149$$

$$\cos 50° = \frac{y}{1.500}$$

$$Y = 1.500 \cos 50° = 0.964 \quad J = 0.964$$

The corrected dimensions are shown in Fig. 14.3(c).
(d) The program is shown in Table 14.2.

TABLE 14.2

N Address	G Code	X Axis	Y Axis	Z Axis	I Arc	J Arc	Feed (in./min)	Spindle Speed (rev/min)	T/H Tool	M Function	Comment
N10	G00/G90	3.0	-3.0	3.0					T01/H01	M06	Load ½-in. end mill spindle
N20								300		M03	Start spindle
N30		0.25	-0.25	0.2							SP to A
N40	G01			-0.25			12.0				Feed
N50			3.25								A to B
N60		-2.25									B to C
N70	G02	-2.392	3.455		0.0	0.75					C to D
N80	G01	-4.69	7.655								D to E
N90	G03	-6.0	8.25		1.149	0.964					E to F
N100	G01	-9.25									F to G
N110			-0.25								G to H
N120		0.25									H to A
N130	G00			3.0							Rapid
N140		3.0	-3.0								To SP
N150										M02	(end of program)

Chap. 14 Questions and Problems

QUESTIONS AND PROBLEMS

14.1. Draw Fig. 14.1(a) through (d) and explain the compensation at each junction point.

14.2. In Fig. 14.1(a) through (d) there are two dashed lines tangent to the cutter. Explain their significance.

14.3. (a) Calculate the dimensions at the junction points in Fig. 14.2 if the drawing outlines a pocket.
(b) Calculate the corrections.
(c) Compensate for the corrections at each junction point.
(d) Write the program in the absolute mode.

14.4. (a) In Fig. 14.4 calculate the X and Y values at each junction point in the absolute mode.
(b) Calculate the correction at each junction point.
(c) Compensate for the end mill shown in Fig. 14.4.
(d) Program the part.

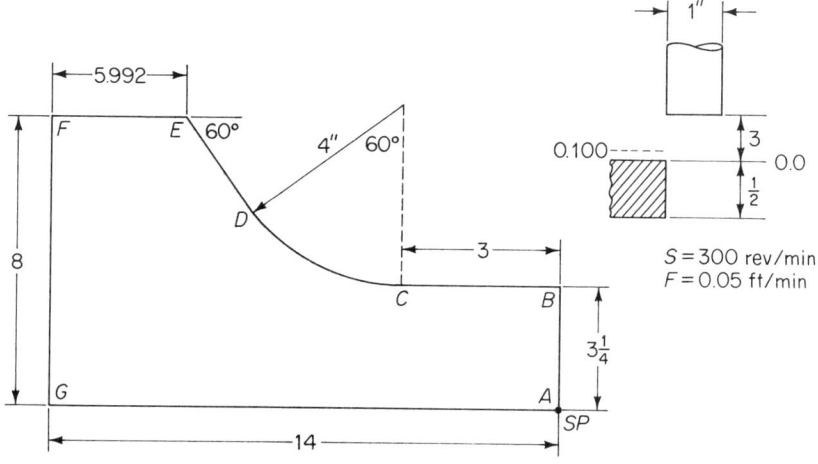

Figure 14.4

14.5. In Fig. 14.4 assume a $\frac{3}{4}$-in.-diameter cutter.
(a) Calculate the X and Y values at each junction point in the absolute mode.
(b) Calculate the corrections.
(c) Make the corrections to the dimensions at each junction point.
(d) Program the part.

14.6. In Fig. 14.5 assume a $\frac{1}{2}$-in.-diameter end mill.
(a) Calculate the X and Y values at each junction.
(b) Calculate the correction factor at each junction.

228   Radius–Angle Combinations   Chap. 14

(c) Calculate the corrected dimension at each junction.
(d) Program the part.

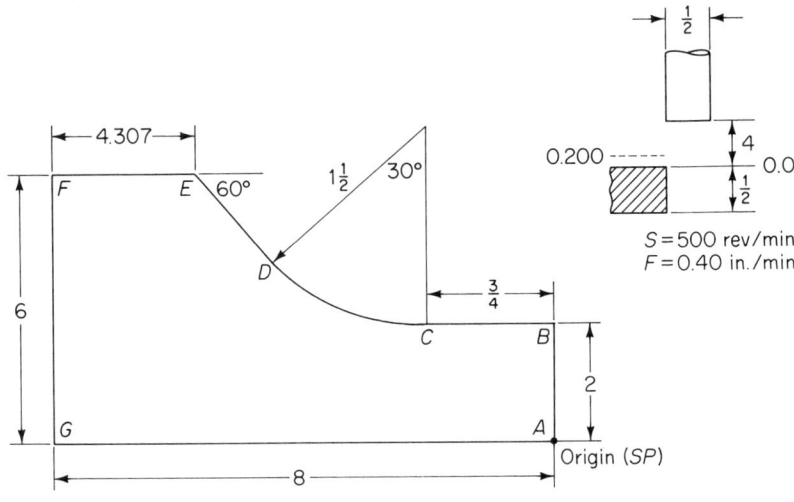

Figure 14.5

14.7. Program Fig. 14.6 in the absolute mode.
  (a) Calculate the X and Y values at each junction.
  (b) Calculate the corrections at each point.
  (c) Calculate the position of the cutter at each point.
  (d) Program the part.

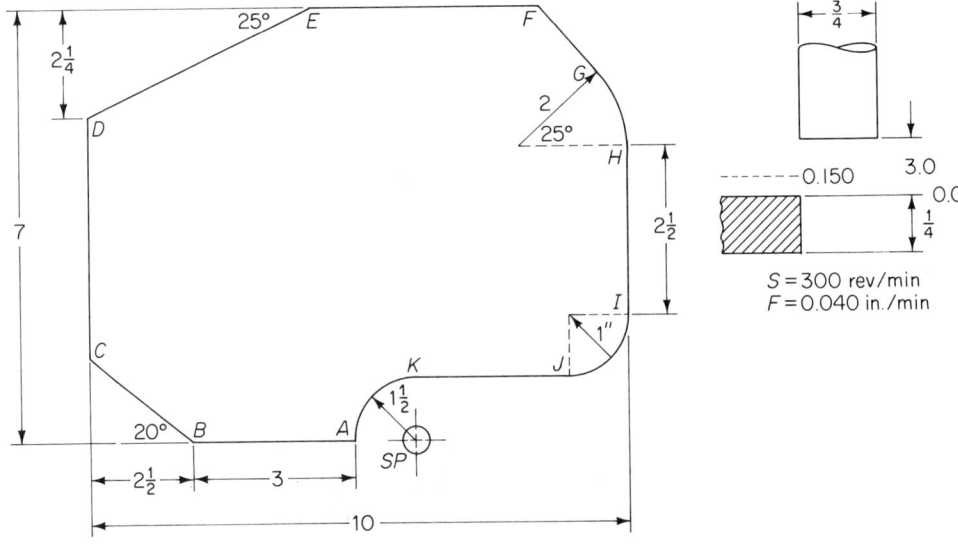

Figure 14.6

14.8. Repeat Prob. 14.7 for the data in Fig. 14.7.

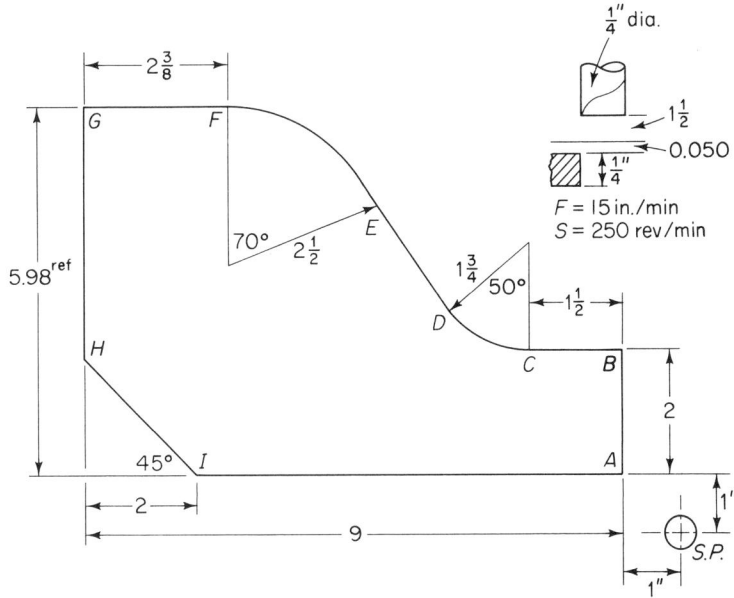

Figure 14.7

14.9. Repeat Prob. 14.7 for the data given in Fig. 14.8.

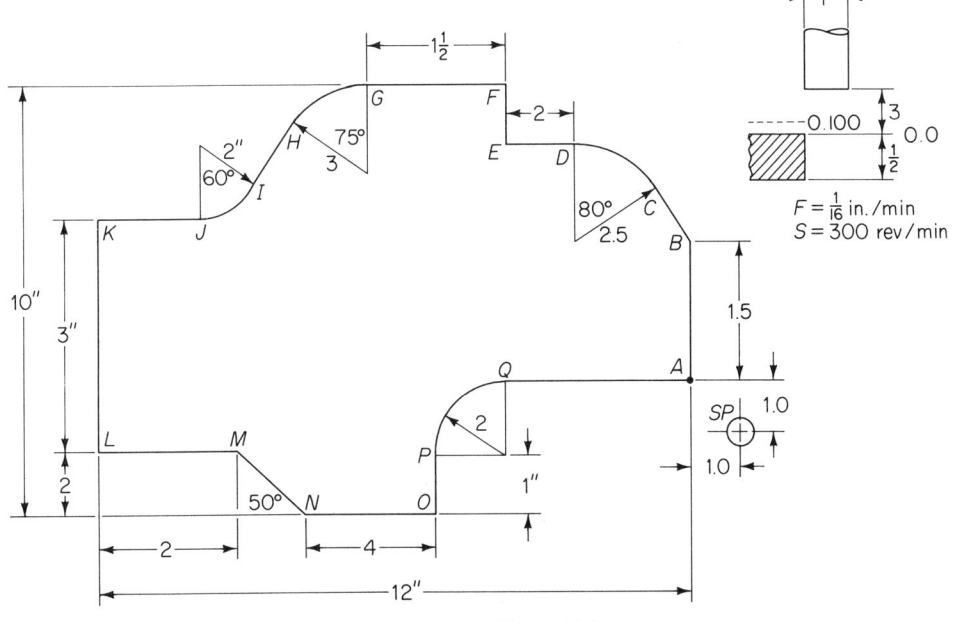

Figure 14.8

14.10. Repeat Prob. 14.7 for the data in Fig. 14.9.

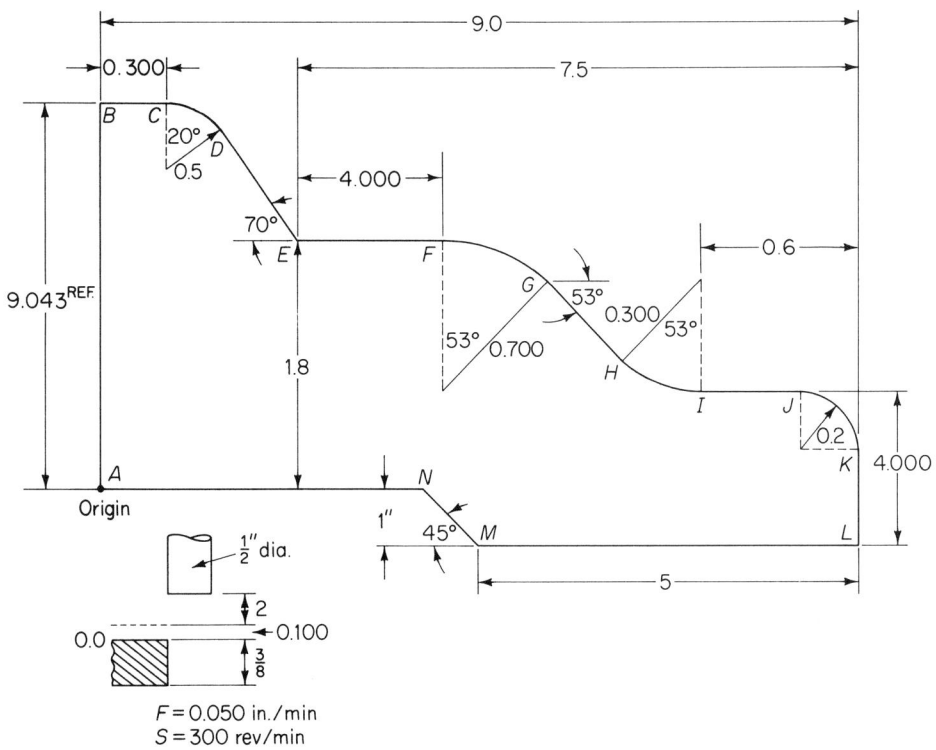

$F = 0.050$ in./min
$S = 300$ rev/min

Figure 14.9

15

Canned Cycles

15.1 CYCLES

Automation requires programmed operations that start at some point, continue through a series of machining operations and through a non-machining cycle, go back to a start point, or proceed to another workpiece that is to be machined. The part to be machined may have a surface larger than the cutter. This would require several passes to machine the entire surface. Machining a pocket into the surface of a workpiece, or drilling several holes into the surface of a stepped workpiece, may dictate cycle-type programs.

These cycle-type programs use a G code. This shortens the need to program each step in a series, or sequence, of operations. Canned cycles may be placed in memory and called up when needed. The main purpose for cycle files is to shorten the programs needed for repetitive operations.

15.2 SURFACE MILLING

Figure 15.1 shows a surface to be machined. A programmer may choose to machine the $6 \times 2\frac{3}{4}$ in. surface by programming each cut separately. This would defeat the advantages that could be gained through cycle programming.

Using cycle programming, the programmer could start the cutter

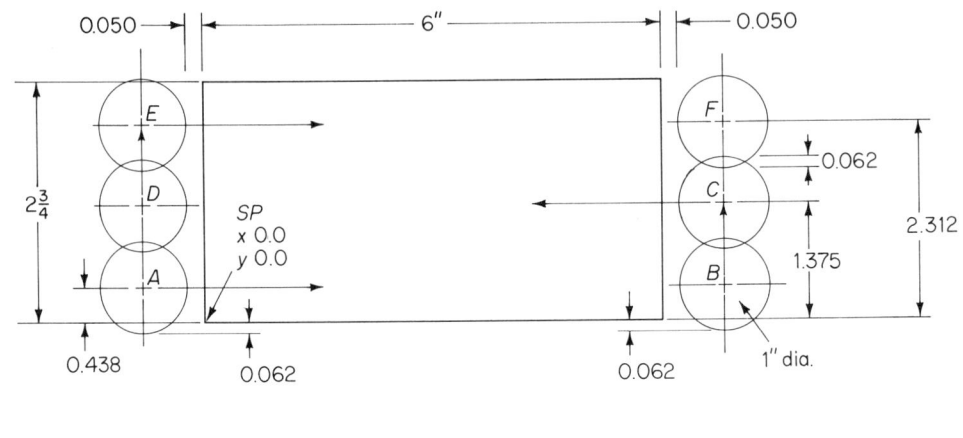
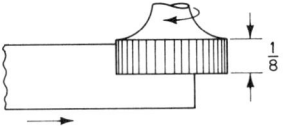

Figure 15.1

off the edge of the surface of the work, cut one length along the positive X axis, move the work, or the cutter, a calculated distance along the Y axis, reverse the cutting operation along the negative X axis, and so on. Repeat the process until the entire surface is machined. It should be noted that the cutter should overlap the 6-in. edge and clear the ends and sides of the workpiece (see Fig. 15.1).

Example 1

The surface of Fig. 15.1 is to be machined with a 1-in.-diameter face mill. The cutter should overlap the X direction of the workpiece by 0.050 in. and the Y direction by 0.062 in. (see Fig. 15.1).

(a) Analyze the movement of the cutter.
(b) Write the conventional program necessary to machine the entire surface of Fig. 15.1 using a feed of 15 in./min. and a cutting speed of 1200 rev/min. Assume a $\frac{1}{8}$-in. depth of cut.
(c) Write the canned cycle program.

Solution In Fig. 15.1 the 1-in.-diameter cutter starts at SP and moves to point A. At A, the cutter moves to a depth of $\frac{1}{8}$ in. below the surface of the work. The movement of the cutter is from A to B, B to C, and so on. Once the cutter reaches point F, the cutter retracts to the home Z position and moves back to position (0, 0) at the start point.

(a) The analysis for the movement of the cutter is as follows: The

Sec. 15.2 Surface Milling

cutter moves from the SP (0, 0) to point A along the X and Y axes, then continues from A to B, B to C, and so on.

SP to A:

$$X = -0.500 + (-0.050) = -0.550 \text{ in.}$$
$$Y = 0.500 - 0.062 = 0.438 \text{ in.}$$

A to B:

$$X = 6.000 + 0.050 + 0.500 = 6.550 \text{ in.}$$
$$Y = 0.438 \text{ in.}$$

B to C:

$$X = 6.650 \text{ in.}$$
$$y = (1.000 + 0.500) - 2(0.0625) = 1.375 \text{ in.}$$

C to D:

$$X = -0.550 \text{ in.}$$
$$Y = 1.375 \text{ in.}$$

D to E:

$$X = -0.550 \text{ in.}$$
$$Y = 1.375 + (1.000 - 0.062) = 2.312 \text{ in.}$$

E to F:

$$X = 6.000 + (0.050 + 0.500) = 6.550 \text{ in.}$$
$$Y = 2.312 \text{ in.}$$

F to SP:

$$X = -0.550 \text{ in.}$$
$$Y = 0.438 \text{ in.}$$

(b) The conventional program is shown in Table 15.1(a).

(c) The canned cycle program is shown in Table 15.1(b). The G77 code in Table 15.1(b) is followed with:

The incremental *movement* along the X axis:

$$6.000 + 2(0.050) + 1.000 = 7.100 \text{ in.}$$

The incremental *movement* along the Y axis:

$$1.375 + 0.500 = 1.875 \text{ in.}$$

The incremental *stepover* along the Y axis:

$$1.000 - 0.0625 = 0.938$$

TABLE 15.1(a)

N Address	G Code	X Axis	Y Axis	Z Axis	Spindle Speed (rev/min)	Feed (in./min)	M Function	Comment
N10					1200		M03	Spindle on: CW
N20	G00	-0.550	0.438	-0.125				Rapid SP to A
N30	G01					15.0		Feed at A
N40		6.550						Feed A to B
N50	G00		1.375					Rapid B to C
N60	G01	-0.550						Feed C to D
N70	G00		2.312					Rapid D to E
N80	G01	6.550						Feed E to F
N90	G00			0.0				Rapid at F
N100		-0.550	0.438					Rapid F to SP
N110							M02	(end of program)

TABLE 15.1(b)

N Address	G Code	X Axis	Y Axis Movement	Y Axis Stepover	Z Axis	Spindle Speed (rev/min)	Feed (in./min)	M Function	Comment
N10						1200		M03	Spindle on: CW
N20	G00	-0.500	0.438						SP to A
N30							15.0		At A
N40	G77	7.100	1.875	0.938	-0.125				Canned cycle
N50					0.0				At F
N60	G00	-0.500	0.438						F to SP
N70								M02	(end of program)

15.3 CANNED CYCLE: MULTIPLE ROW DRILLING

When it becomes necessary to drill a large number of holes in a straight line (or lines), a canned cycle G81 code is used. This may be used together with a G91 incremental or a G90 absolute code.

Example 2

Figure 15.2 shows a series of nine equally spaced drilled holes. They are to be drilled $\frac{1}{2}$ in. apart and 0.100 in. deep. A feed rate of 100 in./min is to be used. Write:

(a) A conventional program in the incremental mode and in the absolute mode.

(b) A canned cycle program in the incremental mode and in the absolute mode.

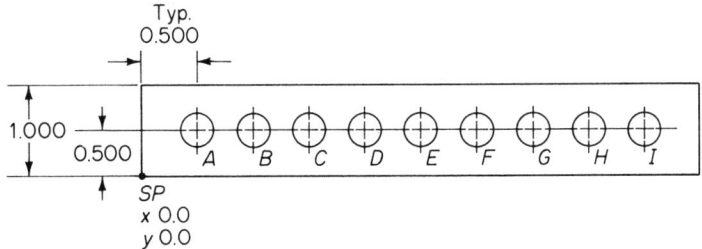

Figure 15.2

Solution (a) The program in the conventional incremental mode is shown in Table 15.2(a).

TABLE 15.2(a)

G Code	X Axis	Y Axis	Z Axis	R Routine	Feed (in./min)	Comment
G91/G00	0.5	0.5				Incremental rapid
G81	0.0	0.0	-0.1	0.5	10.0	At SP, canned cycle
	0.5	0.0				At B
	0.5	0.0				At C
	0.5	0.0				At D
	0.5	0.0				At E
	0.5	0.0				At F
	0.5	0.0				At G
	0.5	0.0				At H
	0.5	0.0				At I

[a]The rapid Z movements, and so on, should be included to complete the program.

The program in the conventional absolute mode is shown in Table 15.2(b).

(b) The incremental canned cycle program for Fig. 15.2 is shown in Table 15.2(c).

The absolute canned cycle program for Fig. 15.2 is shown in Table 15.2(d).

TABLE 15.2(b)

G Code	X Axis	Y Axis	Z Axis	R Routine	Feed (in./min)	Comment
G90/G00	0.5	0.5				
G81	0.5	0.5	-0.1	0.05	10.0	At A
	1.0	0.5				At B
	1.5	0.5				At C
	2.0	0.5				At D
	2.5	0.5				At E
	3.0	0.5				At F
	3.5	0.5				At G
	4.0	0.5				At H
	4.5	0.5				At I

TABLE 15.2(c)[a]

G Code	X_1 Axis	X_2 Axis	Y_1 Axis	Z Axis	R Routine	Feed (in./min)	Comment
G91/G00	0.5		0.5				Increment rapid
G91/G81	4.0	0.5		-0.1	0.05	10.0	Increment canned

[a]G91   indicates the incremental mode
  G81   drill canned cycle
  X_1 4.0   incremental distance A to I
  X_2 0.5   incremental distance between holes
  Z   -0.1   depth of the hole
  R   0.05 routine
  F   10.0 feed (100 in./min)

Sec. 15.4 Canned Cycle: Circular Pocket Milling

TABLE 15.2(d)[a]

G Code	X_1 Axis	X_2 Axis	Y_1 Axis	Z Axis	R Routine	Feed (in./min)	Comment
G90/G00	0.5		0.5				Absolute rapid
G90/G81	4.5	0.5		-0.10	0.05	10.0	Absolute canned

[a] G90   indicates the absolute mode
    G81   drill canned cycle
    X   4.5   absolute distance A to I
    X   0.5   incremental distance between lines
    Z -0.1   depth of hole
    R 0.05   routine
    F 10.0   feed (100 in./min)

15.4 CANNED CYCLE: CIRCULAR POCKET MILLING

Figure 15.3(a) shows a 4.000-in. diameter pocket. The pocket has the starting point and the origin (x = 0, y = 0) at the center, shown in Fig. 15.3(b). The cutter will feed out along a straight line to point A, just short of the required 2-in. radius. This is shown in Fig. 15.3(c). It will then cut an ever-*increasing* radius arc in a counterclockwise direction to point 2. From point 2 it will continue to cut the required 2.000-in. radius in a counterclockwise direction until it returns to point 2. From point 2 it will continue to cut a *decreasing* arc until it reaches point B [Fig. 15.3(c)]. It will complete the cycle by cutting a straight line back to the start point.

Using a smaller and smaller radius, the process above is continued until the pocket has been completed.

Example 3

The circular pocket [Fig. 15.3(a)] is to be milled with a ½-in. two-lip end mill.

    (a) Write the standard program needed to mill the pocket.
    (b) Write the canned cycle program lines needed to mill the pocket. Use a 0.025-in. overlap when machining each successive radius.
    (c) Write the canned cycle program.

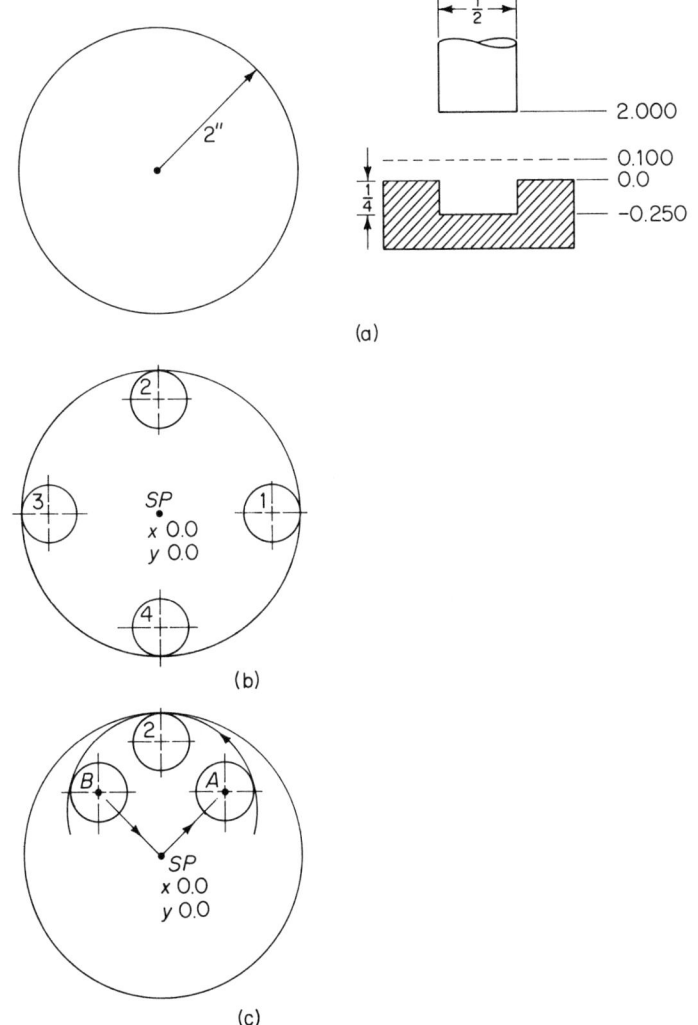

Figure 15.3

Solution (a) The standard program for Fig. 15.3(a). In Fig. 15.3(b) and starting at SP (center of the circle), a straight-line cut is made from SP to 1. Each quadrant is then programmed separately, after which the cutter returns to the SP. The standard program for one cycle is shown in Table 15.3(a).

This process is repeated with successive reduction in radii until the pocket has been completed.

(b) The canned cycle pocket method uses a G79 code. Starting at the center of Fig. 15.3(c), this code causes the cutter to trace a straight-

TABLE 15.3(a)

N Address	G Code	X Axis	Y Axis	Z Axis	I Arc	J Arc	Feed (in./min)	Spindle Speed (rev/min)	T/M Function	Comment
N10	G00	1.75	0.0	0.1					T01/M06	Tool 1
N20	G01			-0.25			10.0	500	M03	Rapid
N30							5.0[a]			SP to 1
N40	G74/G03	0.0	1.75		1.75	0.0				1 to 2
N50	G03	-1.75	0.0		0.0	1.75				2 to 3
N60	G03	0.0	-1.75		1.75	0.0				3 to 4
N70	G03	1.75	0.0		0.0	1.75				4 to 1
N80	G01			0.1						at 1
N90	G00	0.0	0.0	2.0						1 to SP
N100									M02	(end of program)

[a]Plunge feed is one-half the rapid down feed.

TABLE 15.3(b)

Code	Radius	Feed Rate
G79	J1.75	F10

TABLE 15.3(c)

First radius	J =	1.75
Second radius	J =	1.75 − (0.500) + 0.025
	=	2.275
Third radius	J =	1.275 − (0.500) + 0.025
	=	0.800
Fourth radius	J =	0.800 − (0.500) + 0.025
	=	0.325

TABLE 15.3(d)

G79	J1.75	F10	Center radius:	first cut
G79	J1.275			second cut
G79	J0.8			third cut
G79	J0.325			fourth cut

TABLE 15.3(e)

N Address	G Code	X Axis	Y Axis	Z Axis	I Arc	J Arc	Feed (in./min)	Spindle Speed (rev/min)	T/M Function	Comment
N10									T01/M06	Tool 1
N20								500	M03	Rapid
N30	G00	0.0	0.0	0.1						At SP
N40	G01			-0.25			5.0			At 1
N50	G79					1.75	10.0			First circle
N60	G79					1.275	10.0			Second circle
N70	G79					0.8	10.0			Third circle
N80	G79					0.325	10.0			Fourth circle
N90	G01			0.1						Out of hole
N100	G00	0.0	0.0							To SP
N110									M02	(end of program)

Sec. 15.5 Canned Cycle: Pocket Milling

line cut. This radius cut, from SP to A, is somewhat shorter than the required 2-in. radius. The cutter at this point cuts an *increasing* radius arc until it reaches the desired 2-in. radius at point 2.

The cutter will continue to cut the 2-in. radius circle until it has cut a 360° circular groove and has returned to point 2 in Fig. 15.3(c).

From point 2 it will continue to cut a *decreasing* radius arc until it reaches point B in Fig. 15.3(c). At this point it will cut a straight line back to SP. The canned cycle program line is shown in Table 15.3(b).

This process is repeated, with successive shorter radii, until the pocket has been completed. In this case the radii are as shown in Table 15.3(c).

The center cut will overlap the $\frac{1}{2}$-in. plunge cut taken at the SP. The program lines needed to complete the pocket are shown in Table 15.3(d).

(c) The canned cycle program is shown in Table 15.3(e).

15.5 CANNED CYCLE: POCKET MILLING

Assume that a pocket is to be machined into the surface of a workpiece as shown in Fig. 15.4. The procedure would be to start at the center, machine the center pocket, and then move the cutter a calculated distance in the X and Y directions to enlarge the pocket. The enlargement process continues until the desired dimensions are achieved. Once completed, the cutter moves diagonally back to the center of the pocket.

The procedure is to program the outer limits of the pocket, the stepover amounts for each increment, the finishing cut needed to complete the pocket, and the feed rates needed for roughing and finishing.

Thus, in Fig. 15.4, the cutter moves along the inner diagonal to A (note that the letters A, B, C, and so on, indicate the center of the cutter). It will then cut the inner rectangle to complete the small inner pocket. The movements from point A are:

Cut to the left	$-$ X distance
Cut down	$-$ Y distance
Cut to the right	$+$ X distance
Cut up	$+$ Y distance back to point A

The cutter will then move the programmed amount along the diagonal from A to B and complete the next pocket. This process will continue until a finishing cut is required. The cutter will move

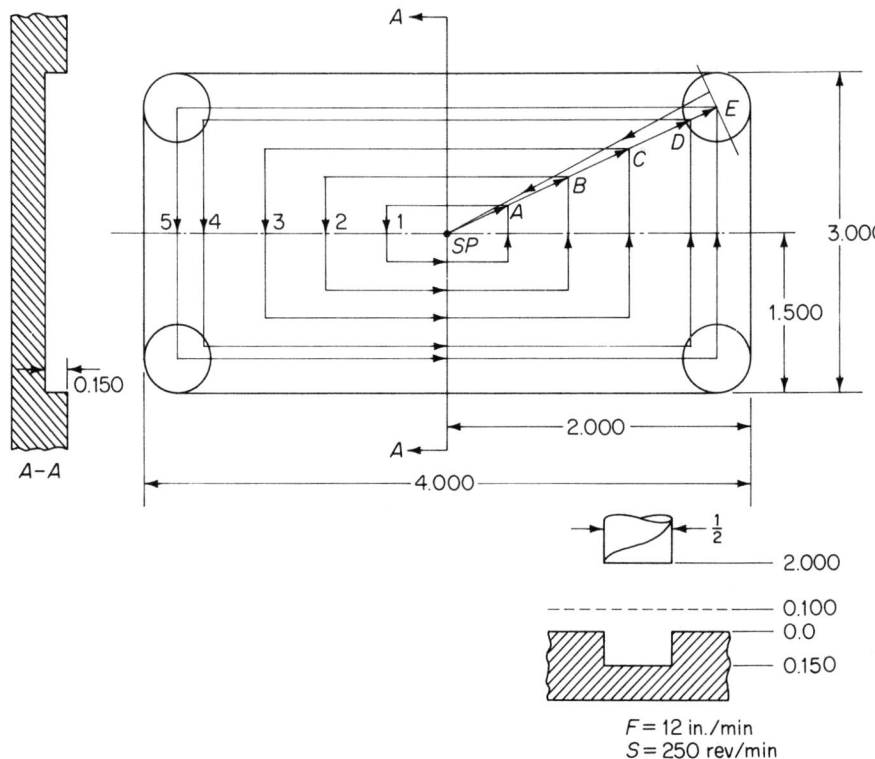

Figure 15.4

the required amount to complete the finishing cut and the pocket. If the *finishing boundary* setover amount is not programmed, the program will automatically default to an amount programmed into the machine memory. Assume that a final boundary cut of 0.015 in. is required. It must be programmed. If it is not included in the program and the default value is 0.030 in., the machine will generate a finishing cut of 0.030 in.

The same holds true for the feed rate needed to take the 225 finishing *boundary cut*. Should the feed rate for the finishing cut not be programmed, the feed rate will default to the *stepover* amount that has been programmed into the machine memory. Assume a finishing cut. The feed rate for the roughing cut is 0.200 in./min. If the feed rate required for finishing is not programmed, the feed rate programmed into the memory of the machine may default to 1.5 times the roughing rate. In this case the finishing feed rate would be 0.300

Sec. 15.5 Canned Cycle: Pocket Milling

in./min. Should a different feed rate be required, it would need to be programmed.

Example 4

Mill the pocket shown in Fig. 15.4 using a $\frac{1}{2}$ in. diameter cutter, a $\frac{1}{4}$ in. setover, a roughing feed of 12 in/min., a finishing feed of 15 in/min. and a spindle speed of 300 rpm.

Solution

TABLE 15.4(a)

G code	G78	Activates canned pocket cycle.
$X_1 n_1$	X1.75	Center of the pocket to the wall: X direction.
$X_2 n_2$	X0.25	Stepover in the X direction. Each successive cut except the last cut.
$X_3 n_3$	X0.015	Finishing boundary cut. If this is not programmed, the default value is activated.
$Y_4 n_4$	Y1.25	Center of the pocket to the wall: Y direction.
$Y_5 n_5$	Y0.180	Stepover in the Y direction. Each successive cut except the last cut. This value is arrive by using the equation $$n_5 = \text{cutter radius}\left(\frac{n_4}{n_1}\right) = 0.25\left(\frac{1.25}{1.75}\right)$$ $$= 0.180 \text{ in.}$$ If this value is not given, the boundary, X and Y, and stepover values will revert to n = 0.25 in.
$F_6 V_6$	F120	The roughing feed of 10 in./min.
$F_7 V_7$	F150	For the boundary cut, the finishing feed of 15 in./min is used. If this value is not programmed, the machine will default to 1.5 × v.
Final cut		The cutter will feed to the center of the pocket to complete the cut.

The pocket program line is shown in Table 15.4(a).

G78 X₁ 1.75 X₂.25 X₃.015 Y₄1.25 Y₅.18 F₆120 F₇150

The program is shown in Table 15.4(b).

TABLE 15.4(b)

N Address	G Code	X_1 Axis	X_2 Axis	X_3 Axis	Y_4 Axis	Y_5 Axis	Z Axis	Spindle Speed (rev/min)	Feed$_6$ (in./min)	Feed$_7$ (in./min)	T/M Function	Comment
N10											T01/M06	Tool 1
N20								300			M03	At SP
N30	G00	0.0			0.0		0.1					At center
N40	G01						-0.15		6.0			Plunge
N50	G78	1.75	0.25	0.015	1.25	0.18			12.0	15.0		Canned cycle
N60	G01						0.1					To 0.100 dp
N70	G00						2.0					Home
N80											M02	(end of program)

15.6 POLAR COORDINATES

It is also possible to write programs in *polar coordinates*. In the polar mode the origin (0,0) is programmed. Then the radius R and the I and J polar coordinates are programmed followed by the incremental *first* angle (30°), Z, and the feed codes. Additional incremental angles (50°) and the XO, YO origin) are programmed.

In Fig. 15.5(a), point H is defined by the angle O, the radius R and a fixed point P. The latter are referred to the x-axis PX. PX is the *polar axis*. If the point H is taken from the polar axis PX in a counterclockwise direction, the angular rotation is designated plus(+). If it is taken clockwise from the polar axis, the angular direction is designated minus(−).

In Fig. 15.5(b)

R is the radius of the bolt circle.

I is the X coordinate from the pole P in the absolute mode.

J is the Y coordinate from the pole P in the absolute mode.

Having defined the point P and the radius, any hole that is on the bolt circle may be determined by stating the angle between that point and the polar axis in either the positive or negative rotational direction.

Example 5

Program Fig. 15.6 in the:

(a) Absolute mode.

(b) Incremental mode.

Solution. (a) The program for Fig. 15.6 is shown in Table 15.5 in the absolute mode. It is important to point out that the second angle (A80) is programmed in the absolute mode. In addition, the direction of rotation is programmed + or −. A30 indicates a counterclockwise rotation because 30 is plus (+). A clockwise rotation would be programmed minus (−), or A-30.

(b) The program in Fig. 15.6 is shown in Table 15.6 in the *incremental mode*. This program is the same as the program in Part (a) to line N30. Beginning with line N40, in the incremental mode (G91), the angles are incremental (A50). Note the return to the absolute mode (G90) in N50.

246 Canned Cycles   Chap. 15

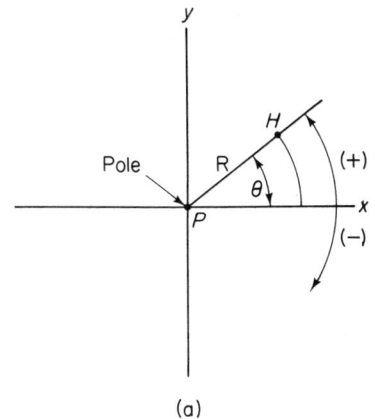

(a)

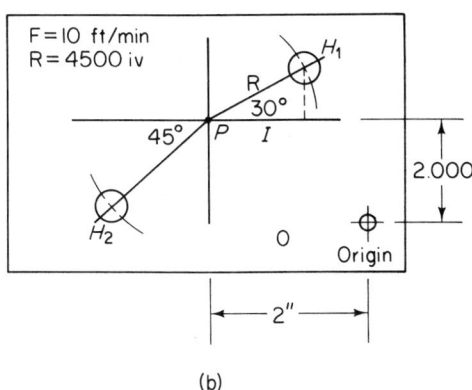

(b)

Figure 15.5

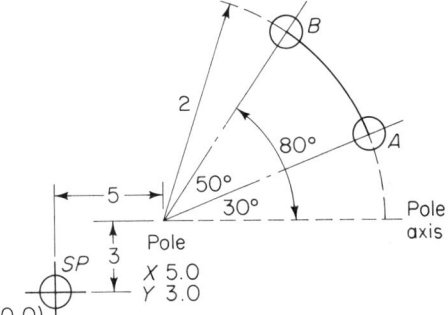

Figure 15.6

Sec. 15.6 Polar Coordinates

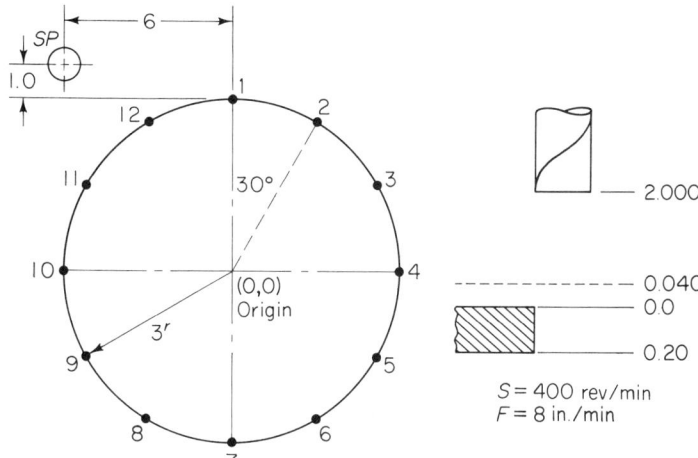

Figure 15.7

Example 6

Program Fig. 15.7 in the polar system in the *absolute mode*.

Solution The program is shown in Table 15.7 in the polar mode.

TABLE 15.5

N Address	Codes	X Axis	Y Axis	Z Axis	I Arc	J Arc	Feed (in./min)	Comment
N10	G00 G90	0.0	0.0	0.05				At SP
N20	R2				5.0	3.0		At pole
N30	G81 A30			0.1			100.0	(point 1)
N40	A80 (50°+30°)							(point 2)
N50	G00 A80	0.0	0.0					

G91 indicates the *absolute mode*
X0, Y0 is the origin
I and J the coordinates of the polar axis from which the radius is generated
G81 the drill cycle (holes A and B)
A30 indicates the first angular movement
A80 indicates the second angular movement
G00 back to the origin

TABLE 15.6

N Address	Codes	X Axis	Y Axis	Z Axis	I Arc	J Arc	Feed (in./min)
N10	G00 G90	0.0	0.0	0.05			
N20	R2.0				5.0	3.0	
N30	G81 A30			0.1			100.0
N40	G91 A50						
N50	G00 G90	0.0	0.0				

TABLE 15.7

N Address	Codes	X Axis	Y Axis	Z Axis	I Arc	J Arc	Feed (in./min)
N10	G00 G90	-6.0	4.0				
N20	R3. A90			0.1	0.0	0.0	
N30	G81 A30 A60			0.5			150.0
N40	G00	-6.0	1.0				

QUESTIONS AND PROBLEMS

15.1. Why are canned cycles useful to a programmer?

15.2. Using conventional programming methods, write the absolute program for Fig. 15.8. You are to use a 3.000-in. face mill. Each cut should overlap the preceding cut and the edge of the work by 0.050 in.

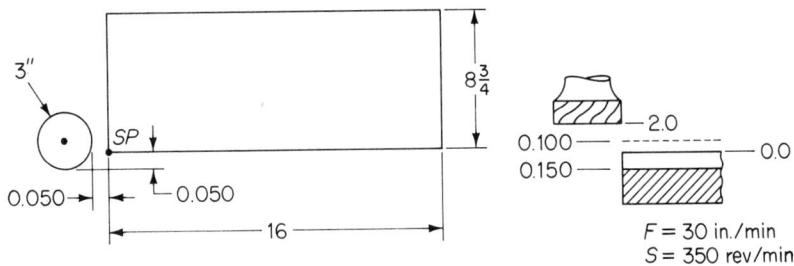

Figure 15.8

15.3. Using a canned cycle, repeat Prob. 15.2.

15.4. Write a conventional program for Fig. 15.9. Use the absolute mode of programming. The first cut should cut a groove in surface *A* and machine face *B*. Use a 2-in. face milling cutter, a feed of 15 in./min, and a spindle speed of 200 rev/min.

15.5. Using a canned cycle, write the program for Fig. 15.9. Use the data from Prob. 15.4.

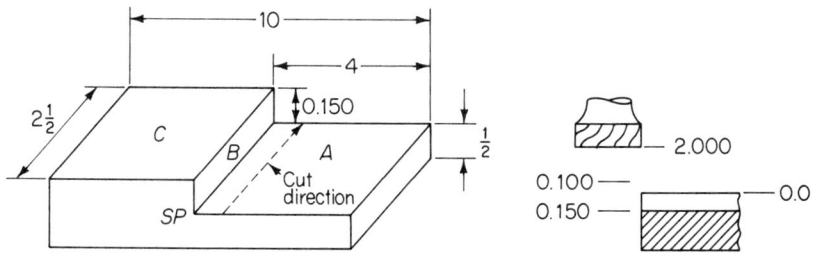

Figure 15.9

15.6. Five equally spaced holes are to be drilled as shown in Fig. 15.10.
   (a) Program the drilling operation in the conventional incremental mode.
   (b) Using an incremental canned cycle, program the drilling operation in Fig. 15.10.

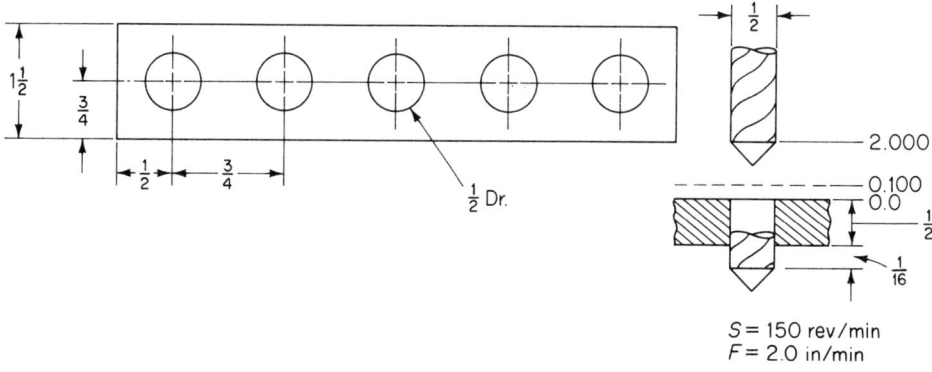

Figure 15.10

15.7. Trace the movement of the cutter in Fig. 15.3 when a canned cycle is used to program the circular pocket. Explain each move.

15.8. Assume the same conditions as those shown in Fig. 15.3 except that a 1-in. groove is to be machined with an end mill.
   (a) Using conventional programming, write the program needed to cut a 1-in. circular groove.
   (b) Repeat part (a) using a canned cycle.

15.9. Using a 1-in.-diameter end mill, cut the pocket in Fig. 15.3. Use the data from Prob. 15.8.
   (a) Use conventional programming. G74
   (b) Use a pocket cycle file. G75
   (c) Use G79 internal pocket.

15.10. Analyze, describe, and document the movement of the cutter in Fig. 15.4.

250 Canned Cycles   Chap. 15

15.11. (a) Define stepover.
  (b) Define boundary setover.
  (c) Explain what happens if a boundary cut is not programmed.
  (d) Repeat Prob. 15.11(c) if the feedrate is not programmed.

15.12. Given a canned cycle file

  G78X4.7X.4X.02Y2.5Y.15F200F260

  (a) Explain each of the terms in this file program.
  (b) Assume that X.02 is not programmed. What would the default value be?
  (c) Repeat part (b) if Y.15 is not programmed.
  (d) Assume that F260 is not programmed. What would the default value be?

15.13. Explain the cutter movement in Fig. 15.4.

15.14. Using conventional absolute programming, program the pocket shown in Fig. 15.11.

15.15. Using a pocket cycle file, program Fig. 15.11.

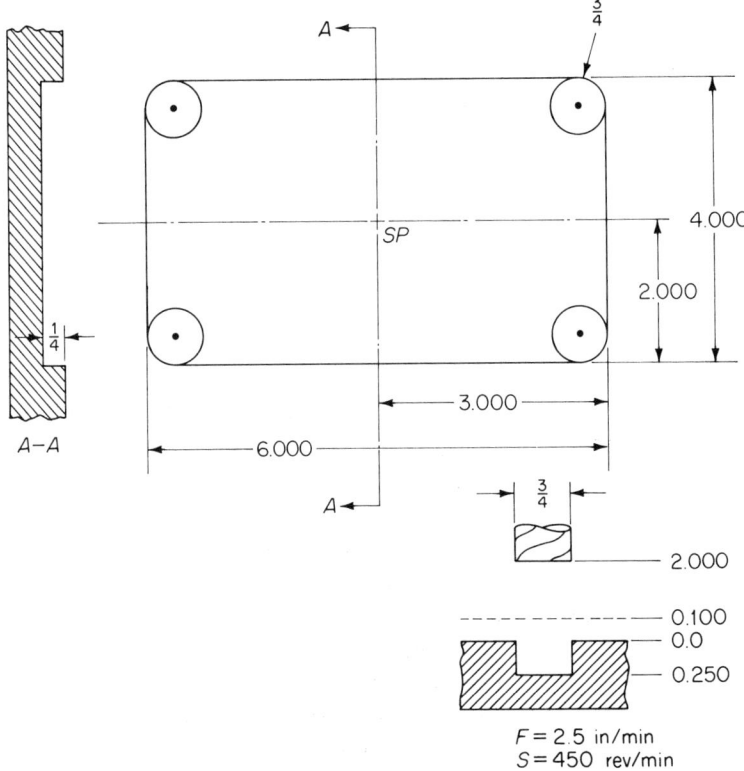

Figure 15.11

15.16. Using polar coordinates, program Fig. 15.12. Show the movement in a clockwise direction from point A to G.

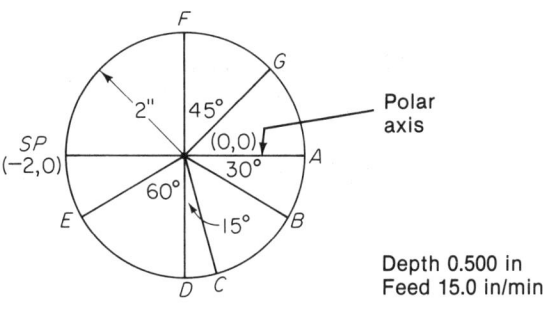

Figure 15.12

Depth 0.500 in
Feed 15.0 in/min

16

The Lathe: Radius Compensation

16.1 LINEAR CONTOURING

Radius compensation on a lathe is calculated as a *function of the surface to be cut*. Correction on the milling machine was applied to the center of the cutter at the intersection of two surfaces. Correction on the lathe is applied to the contact point of the cutter with the work. The student should compare Fig. 14.1 with Fig. 16.1(a).

At point A in Fig. 16.1(a), the distance

$$x = r$$

and

$$\tan \theta° = \frac{c}{r}$$

c = compensation
k = correction factor
r = radius of cutter

Thus

$$c = r \tan \theta°$$

The c distance is the compensation. The correction factor is

$$k = r - c$$

Sec. 16.1   Linear Contouring

At point B in Fig. 16.1(a),

$$y = r$$

and

$$\tan \theta° = \frac{c}{r}$$

$$c = r \tan \theta°$$

The distance c is the compensation and the correction factor is

$$k = r - c$$

It is important that the student *not* rely on the equations. An analysis must be done at each junction point on the drawing. A complete study of Fig. 16.1(b) through (d) should be made. *It should be noted that before correction (dashed lines), the cutter was tangent to the dimension leader lines in the X and Y directions. After the correction factor has been applied to the dimension lines, the tool nose radius (TNR) is tangent both to the surface that has just been completed and to the surface to be cut.*

Example 1

Figure 16.2(a) shows a drawing of a stepped cylinder that is to be machined. Assume that a $\frac{1}{16}$-in. tool nose radius is to be used. The program is to be in the absolute mode.

(a) Calculate the X and Y values at each junction point.
(b) Calculate the correction factors and apply them to the dimensions as calculated in Fig. 16.2(c).
(c) Make a new drawing and insert these corrected dimensions on a drawing.
(d) Using a cutting speed of 200 rev/min, a feed of 0.020 in./rev, and the information developed in part (c), write the program for Fig. 16.2.

Solution   If cutting takes place with the tool in front of the work as in a conventional lathe, the drawings appear as in Fig. 16.2(a). If cutting takes place with the tool in back of the work, the drawing appears as shown in Fig. 16.2(b). This must be checked out before programming is possible.

It should be noted that all movements parallel to the centerline of the work are Z movements. Those perpendicular to the centerline of the work are X movements. Normally, cutting takes place from right to left. Right-to-left movement of the tool is programmed minus ($-$).

254 The Lathe: Radius Compensation   Chap. 16

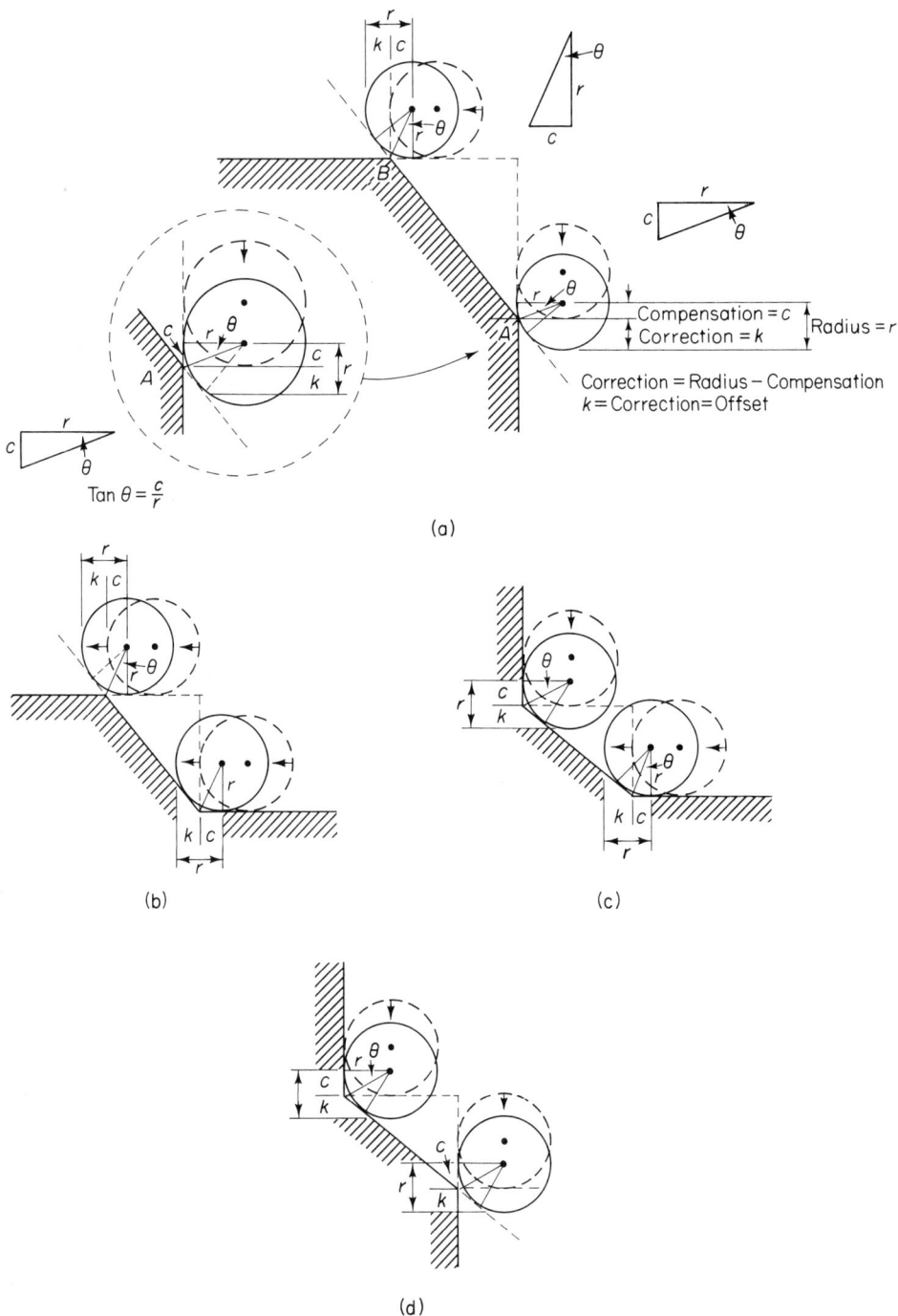

Figure 16.1

Sec. 16.1 Linear Contouring

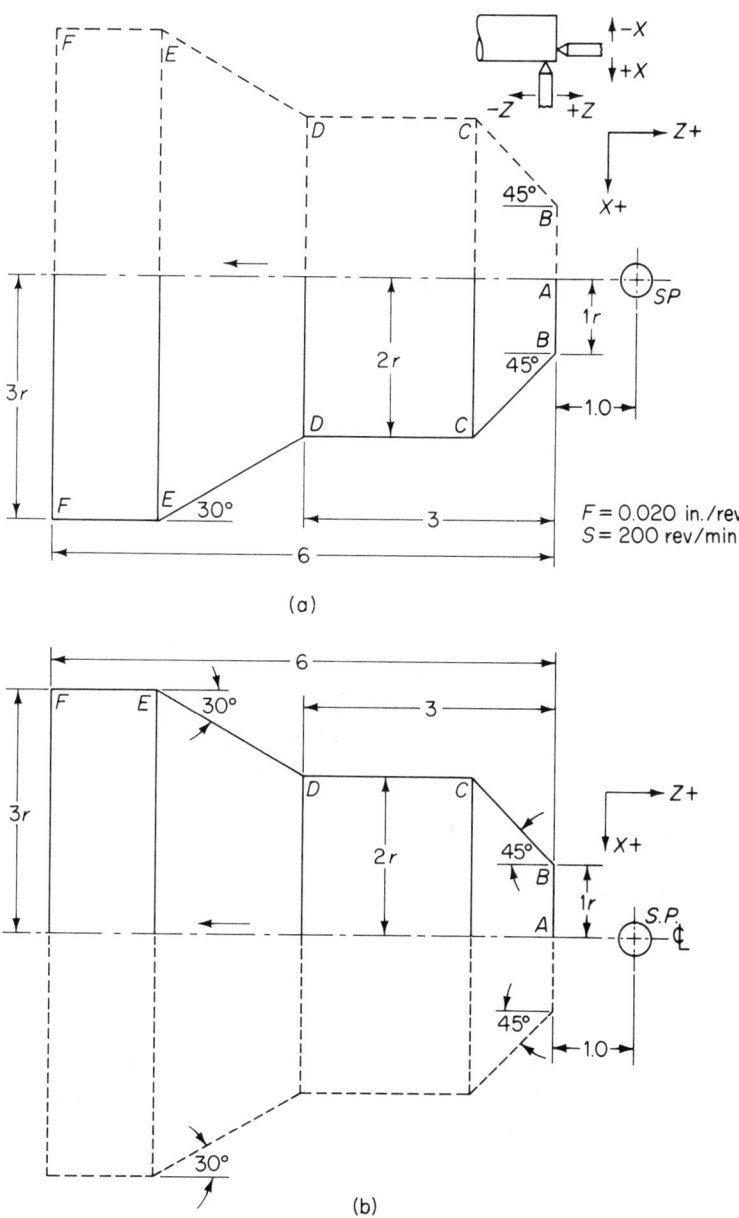

Figure 16.2

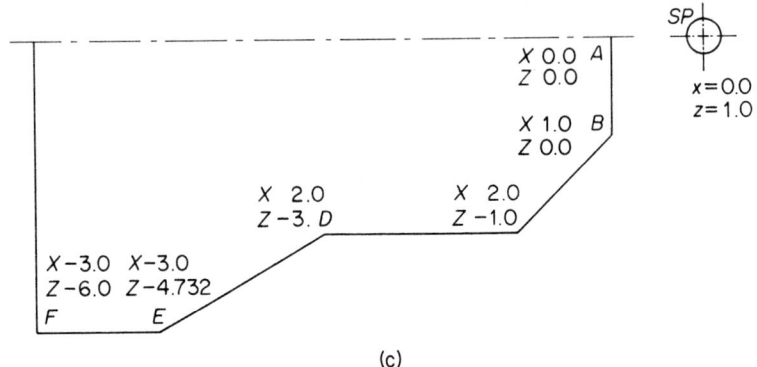

(c)

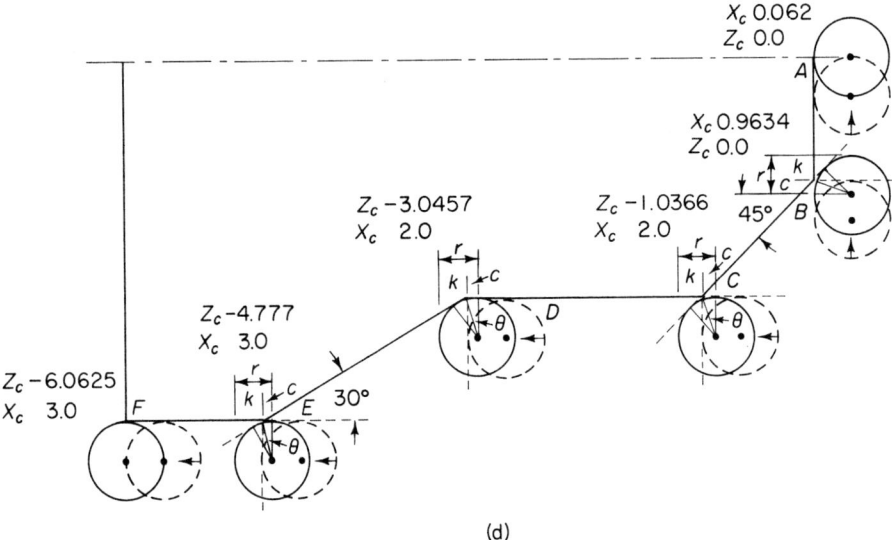

(d)

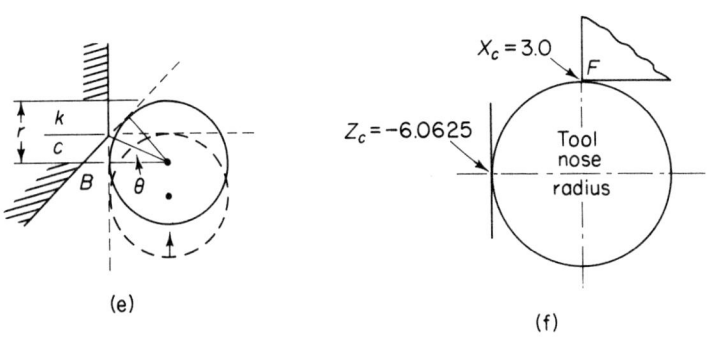

(e)

(f)

Figure 16.2 (Continued)

Sec. 16.1 Linear Contouring

Movement of the tool toward the operator is programmed plus (+0) [see Fig. 16.2(a)].

(a) The workpiece is first dimensioned as though there were no TNR compensation. This is shown in Fig. 16.2(c).

At A:

$$X = 0.000$$
$$Y = 0.000$$

A to B:

$$X = 0.000 + 1.000 = 1.000$$
$$Z = 0.000$$

B to C:

$$X = 1.000 + 1.000 = 2.000$$
$$Z = 0.000 - 1.000 = -1.000$$

C to D:

$$X = 2.000$$
$$Z = -1.000 - 2.000 = -3.000$$

D to E:

$$X = 2.000 + 1.000 = 3.000$$
$$Z = -3.000 - 1.732 = -4.732$$

$$z = \frac{1.000}{\tan 30°} = 1.732$$

E to F:

$$X = 3.000$$
$$Z = -4.732 - (6.000 - 3.000 - 1.732) = -6.000$$

F to A:

$$X = 3.000 - 3.000 = 0.000$$
$$Z = 6.000 - 6.000 = 0.000$$

(b) The corrected X and Z values are as follows. The tool nose radius (TNR) is to be $\frac{1}{16}$ in. Corrections should be applied to Fig. 16.2(c) and inserted into Fig. 16.2(d).

At A:

$$X = 0.062$$
$$Z = 0.000$$

At B [see Fig. 16.2(e)] the compensation is

$$\theta° = 90° - \frac{180° - 45°}{2} = 22.5°$$

$$c = 0.0625 \tan 22.5° = 0.0259$$

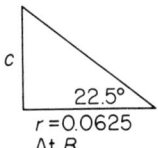

At B

The correction factor is

$$k = 0.0625 - 0.0259 = 0.0366$$

The dimensions at B [Fig. 16.2(e)] at the cutting edge are

$$X_c = 1.000 \text{ [from Fig. 16.21c]} - 0.0366 = 0.9634$$

$$Z_c = 0.000$$

At C compensation is

$$\theta° = 90° - \frac{180° - 45°}{2} = 22.5°$$

$$c = 0.0625 \tan 22.5° = 0.0259$$

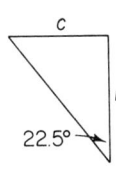

At C

The correction factor is

$$k = 0.0625 - 0.0259 = 0.0366$$

The dimensions at C are

$$X_c = 2.000$$

$$Z_c = -1.000 - 0.0366 = -1.0366$$

At D the compensation is

$$\theta° = 90° - \frac{180° - 30°}{2} = 15°$$

$$c = 0.0625 \tan 15° = 0.0167$$

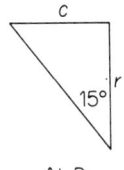

At D

The correction factor is

$$k = 0.0625 - 0.0167 = 0.0457$$

The dimensions at D are

$$X_c = 2.000$$

$$Z_c = -3.000 - 0.0457 = -3.0457$$

TABLE 16.1

N Address	G Code	X Axis	Z Axis	Spindle Speed (rev/min)	Feed (in./rev)	M Function	Comment
N10				200		M04	Spindle on
N20	G00	0.0	1.0				Rapid
N30	G01	0.082	0.0		0.02		Start to A
N40		0.9634					A to B
N50		2.0	-1.0366				B to C
N60			-3.0457				C to D
N70		3.0	-4.777				D to E
N80			-6.0625				E to F
N90	G00	3.1					Rapid off work
N100			1.0				Back to start
N110		0.0					SP
N120						M02	(end of program)

At E the compensation is

$$\theta° = 90° - \frac{180° - 30°}{2} = 15°$$

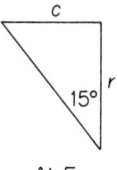

At E

$$c = 0.0625 \tan 15° = 0.0167$$

The correction factor is

$$k = 0.0625 - 0.0167 = 0.0457$$

The dimensions at E are

$$X_c = 3.000$$

$$Z_c = -4.732 - 0.0457 = -4.777$$

At F [Fig. 16.2(f)]

$$X_c = 3.000$$

$$Z_c = -6.000 - 0.0625 = -6.0625$$

In contrast, the milling cutter, which is dimensioned at the centerline of the cutter; the radius of a lathe tool bit [Fig. 16.2(f)] is dimensioned at the cutting edge.

(c) The program is shown in Table 16.1.

16.2 CIRCULAR INTERPOLATION

Circular interpolation, coupled with tool nose radius offset, is illustrated in Fig. 16.3. *The nose radii are positioned tangent to the dimension lines.* The corrections are made as shown.

Example 2

Program Fig. 16.4(a) in the absolute mode if the part is to be machined on a lathe using a ¼-in. radius tool bit.

(a) Dimension a drawing as though there were *no* TNR.
(b) Apply the TNR dimensions to the drawing.
(c) Write the program assuming a feed of 0.015 in./rev and a spindle speed of 200 rev/min.

Solution (a) The dimensions at the junction points are shown in Fig. 16.4(b).

At A:

$$X = 0.000$$
$$Z = 0.000$$

At B:

$$X = 0.000 + 1.000 = 1.000$$
$$Z = 0.000$$

Sec. 16.2 Circular Interpolation

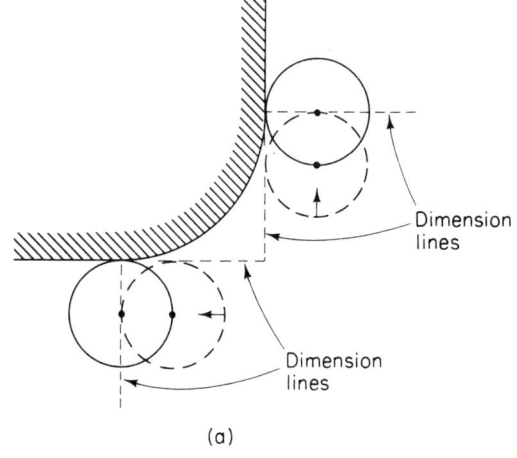

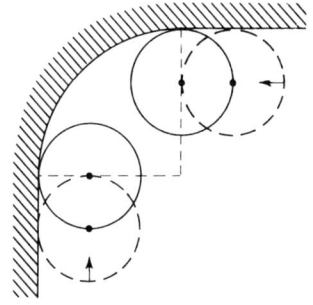

(b)   Figure 16.3

B to C:

$\quad$ X = 1.000 + 0.500

$\quad\quad$ = 1.500

$\quad$ Z = 0.000 − 0.500

$\quad\quad$ = −0.500

C to D:

$\quad$ X = 1.500

$\quad$ Z = −0.500 − 1.875

$\quad\quad$ = −2.375

D to E:

$\quad$ X = 1.500 + 0.625

$\quad\quad$ = 2.125

$\quad$ Z = −2.375 − 0.625

$\quad\quad$ = −3.000

E to F:

$\quad$ X = 2.125 + 1.375

$\quad\quad$ = 3.500

$\quad$ Z = −3.000

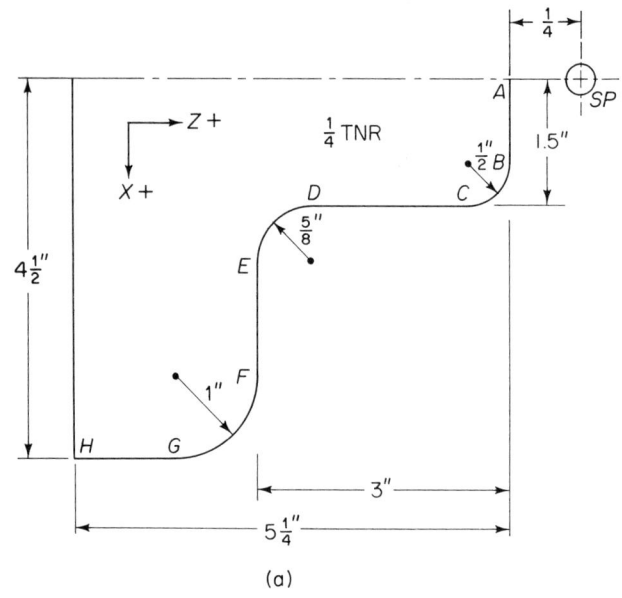

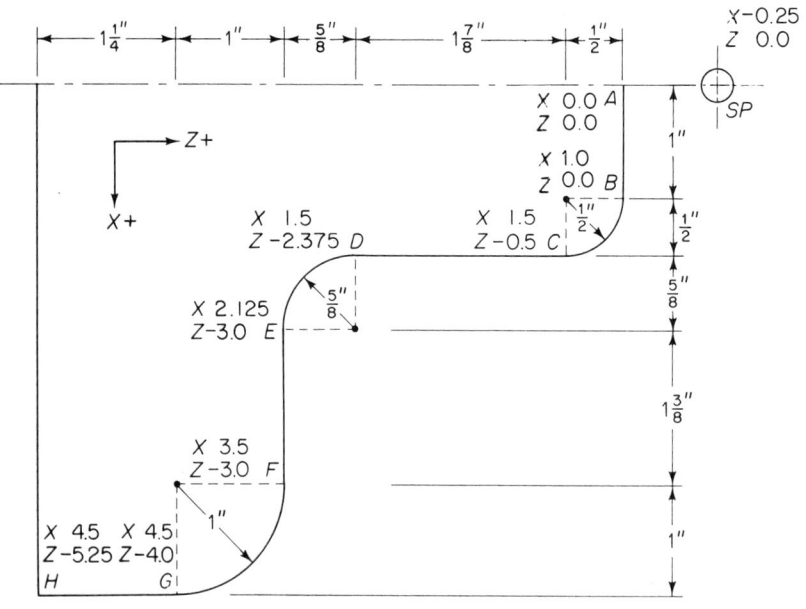

Figure 16.4

Sec. 16.2   Circular Interpolation   263

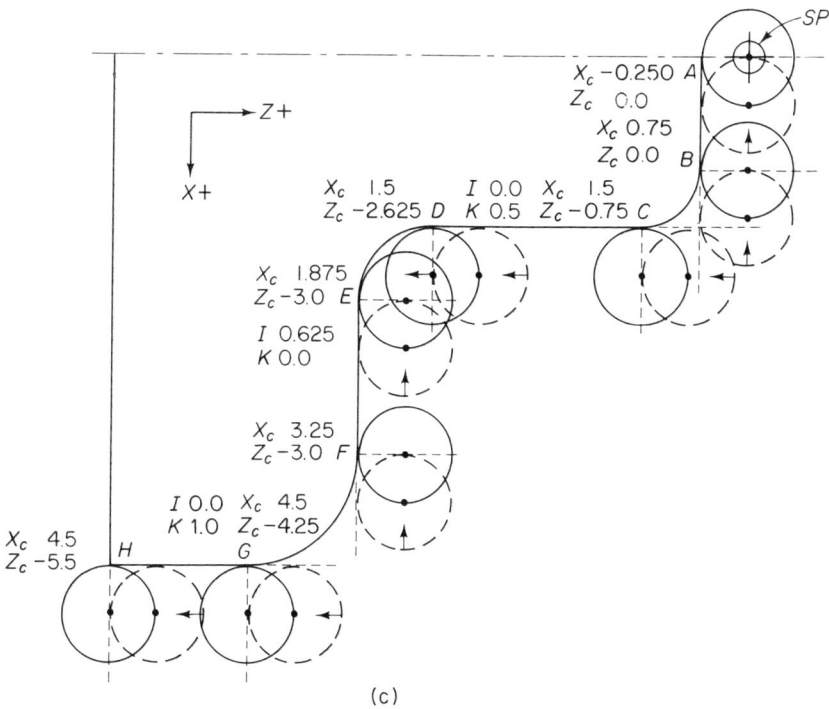

Figure 16.4 (Continued)

F to G:

X = 3.500 + 1.000

    = 4.500

Z = −3.000 − 1.000

    = −4.000

H to A:

X = 4.500 − 4.500

    = 0.000

Z = −5.250 + 5.250

    = 0.000

G to H:

X = 4.500

Z = −4.000 − 1.250

    = −5.250

(b) The next step is to provide the tool nose radius compensation of $\frac{1}{4}$ in. The X and Z dimensions are taken from Fig. 16.4(b), corrected,

TABLE 16.2

N Address	G Code	X Axis	Z Axis	I Arc	K Arc	Spindle Speed (rev/min)	Feed (in./rev)	M Function	Comment
N10						200		M04	Turn spindle CCW
N20	G01	-0.25	0.0				0.015		To A
N30		0.75							A to B
N40	G02	1.5	-0.75	0.0	0.05				B to C
N50	G01		-2.625						C to D
N60	G03	1.875	-3.0	0.625	0.0				D to E
N70	G01	3.25							E to F
N80	G02	4.5	-4.25	0.0	1.0				F to G
N90	G01		-5.5						G to H
N100	G00	4.6							Off the work
N110		-0.25	0.0						To start position
N120								M02	(end of program)

Sec. 16.3   Radius–Angle Combinations

At A:
  $X = 0.250$
  $Z = 0.000$

A to B:
  $X = 1.000 - 0.250$
  $ = 0.750$
  $Z = 0.000$

B to C:
  $X = 1.500$
  $Z = -0.500 - 0.250$
  $ = -0.750$
  $I = 0.0$
  $K = 0.05$

C to D:
  $X = 1.500$
  $Z = -2.375 - 0.250$
  $ = -2.625$

D to E:
  $X = 2.125 - 0.250$
  $ = 1.875$
  $Z = -3.000$
  $I = 0.625$
  $K = 0.0$

E to F:
  $X = 3.500 - 0.250$
  $ = 3.250$
  $Z = -3.000$

F to G:
  $X = 4.500$
  $Z = -4.000 - 0.250$
  $ = -4.250$
  $I = 0.0$
  $K = 1.0$

G to H:
  $X = 4.500$
  $Z = -5.250 - 0.250$
  $ = -5.500$

H to A:
  $X = -0.250$
  $Z = 0.000$

(c) The program is shown in Table 16.2.

16.3 RADIUS–ANGLE COMBINATIONS

In Fig. 16.5(a) the tool bit is to trace a radius R and blend into the cut angle $\theta°$. The tool nose radius (TNR), centered at A, is shown *tangent to the dimension lines*. The X and Z dimensions have *not*

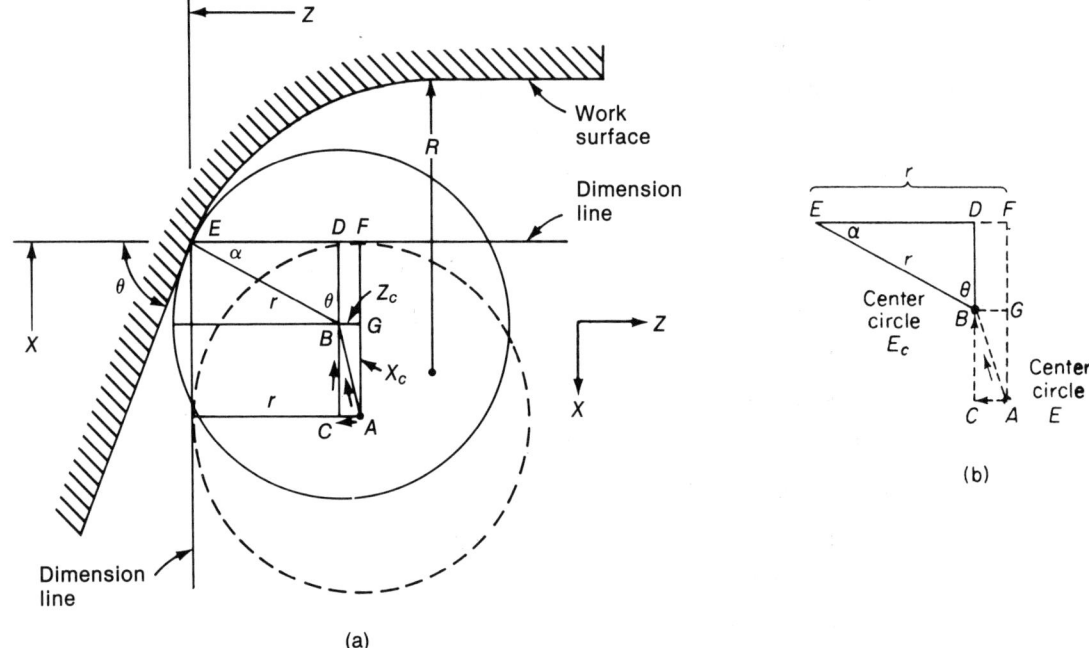

Figure 16.5

been corrected. The TNR must be moved so that its center is at B. This will place the at the point of tangency E of the radius of the work and the angle to be cut. The compensation triangle ABC is shown in Fig. 16.5(b). The compensation distance are calculated in triangle EBD. In Fig. 16.5(b),

$$ED = r \sin \theta°$$

and

$$Bd = r \cos \theta°$$

Since $EF = r$,

$$CA = DF = EF - ED$$

and because $DC = r$,

$$BC = DC - DB$$

Example 3

Calculate the compensation triangle and the corrections in Fig. 16.6(a).

Solution In Fig. 16.6(b),

$$c_x = BD = 0.250 \cos 60° = 0.125$$

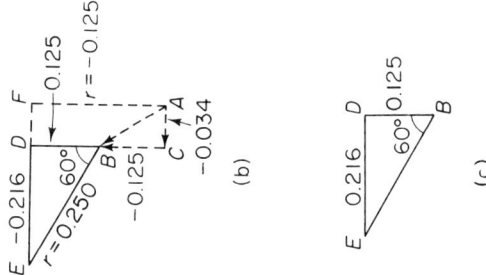
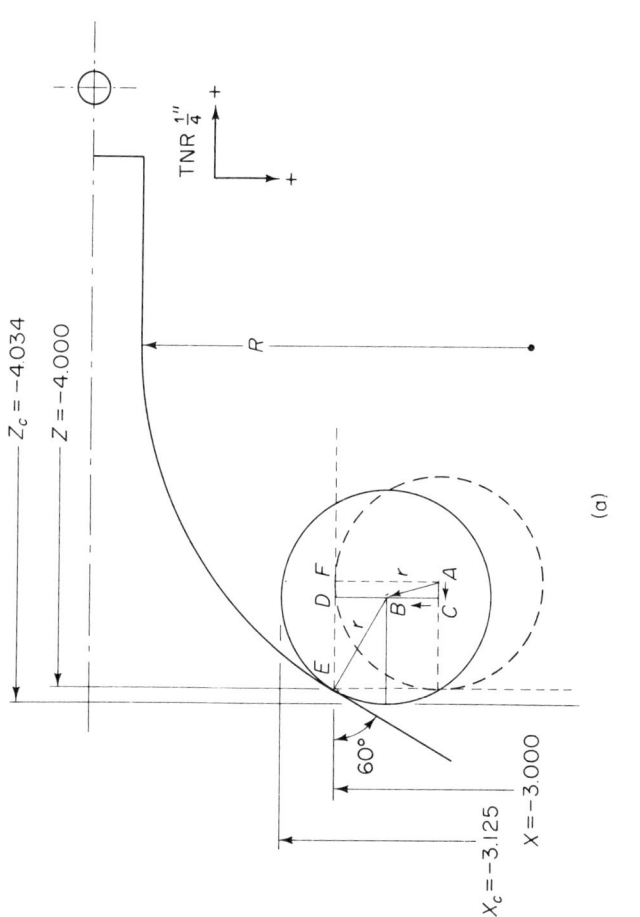

Figure 16.6

Since $CB = k_x$,
$$k_x = 0.250 - 0.125 = 0.125$$
Also,
$$c_z = ED = 0.250 \sin 60° = -0.216$$
Since $CA = k_z$,
$$k_z = -0.250 + 0.216 = -0.034$$

Example 4

Given a $\frac{1}{4}$-in. tool nose radius.

(a) Dimension Fig. 16.7(a) in the absolute mode. Do not apply TNR compensation.
(b) Calculate the TNR compensations and apply the correct dimensions to the drawing.
(c) Program the part using a spindle speed of 150 rev/min and a feed of 0.010 in./rev.

Solution (a) The drawing is redimensioned so that the absolute mode may be calculated. The redimensioned drawing is shown in Fig. 16.7(b) and (d).
In Fig. 16.7(c) in triangle HID,

$$\sin 40° = \frac{HI}{1.250}$$

$$HI = 1.250 \cos 40° = 0.803$$

$$IE = HE - HI = 1.250 - 0.803 = 0.447$$

$$ID = 1.250 \cos 40° = 0.958$$

In triangle CDG,

$$\tan 50° = \frac{GD}{1.513}$$

$$GD = 1.513 \tan 50° = 1.803$$

Also,

$$GD = 4.000 - (0.750 + 1.000 + 0.447) = 1.803$$

Therefore,

$$CB = 2.250 - 1.000 = 1.250$$

and

$$EF = 5.500 - (1.000 + 1.250 + 1.513 + 0.958) = 0.779$$

Sec. 16.3 Radius–Angle Combinations

The absolute dimensions are as follows:

At ⊕:
  X = −0.250
  Z = 0.000

At A:
  X = 0.750
  Z = 0.000

A to B:
  X = 0.750 + 1.000
    = 1.750
  Z = 0.000 − 1.000
    = −1.000

B to C:
  X = 1.750
  Z = −1.000 − 1.250
    = −2.250

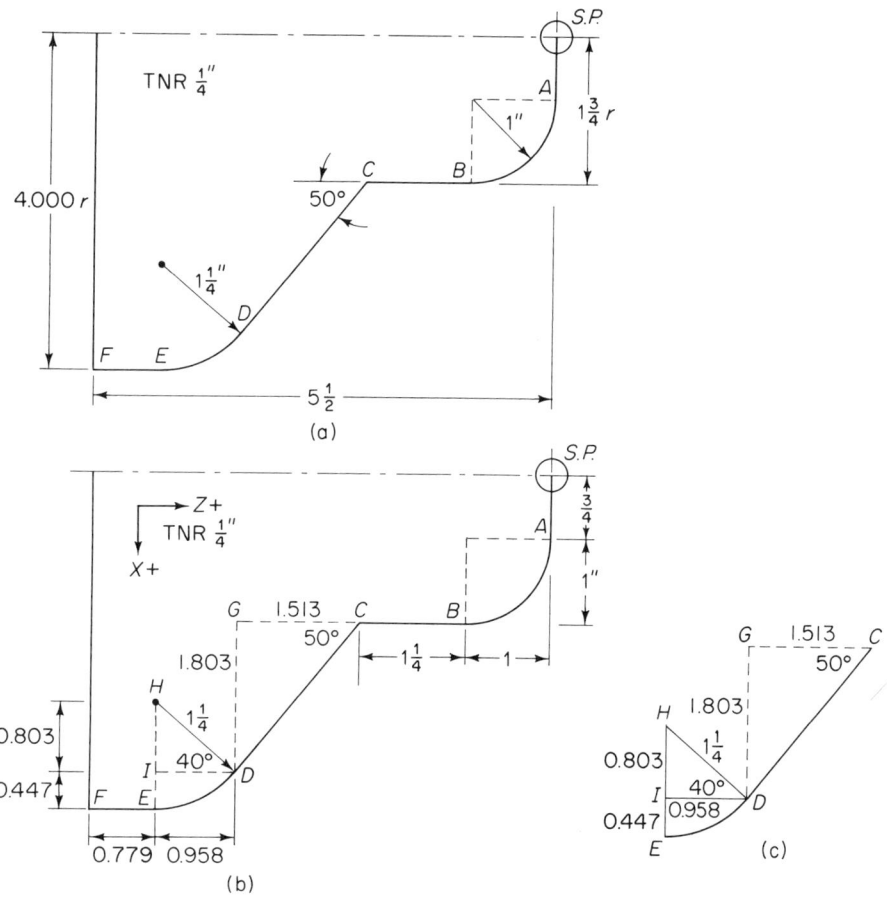

Figure 16.7

(Continued)

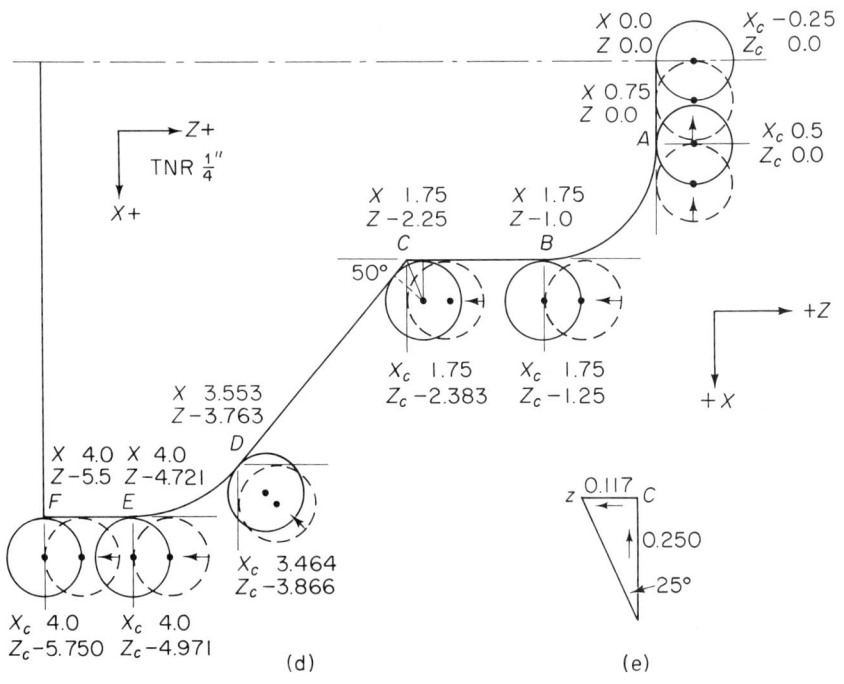

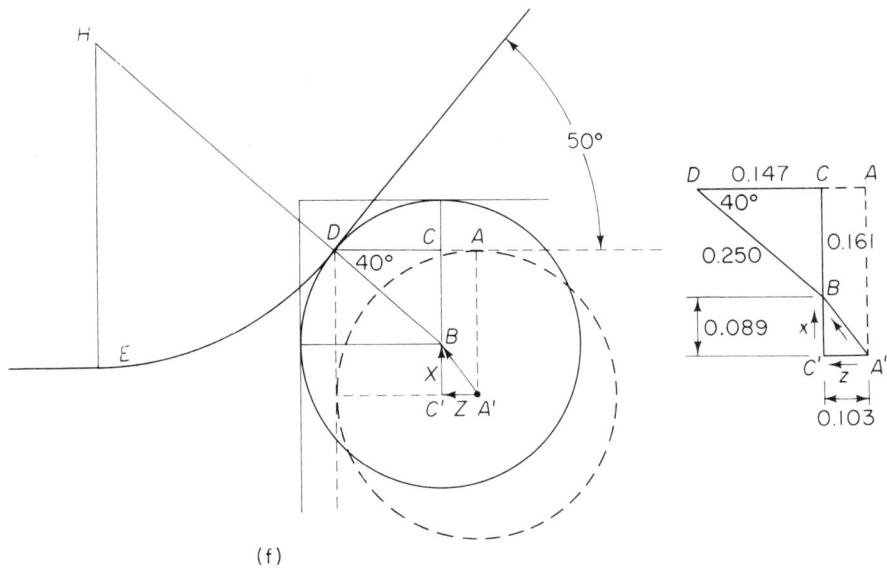

Figure 16.7 (Continued)

Sec. 16.3   Radius–Angle Combinations

C to D:
$$X = 1.750 + 1.803$$
$$= 3.553$$
$$Z = -2.250 - 1.513$$
$$= -3.763$$

D to E:
$$X = 3.553 + 0.447$$
$$= 4.000$$
$$Z = -3.763 - 0.958$$
$$= -4.721$$

D to E:
$$X = 4.000$$
$$Z = -4.721 - 0.779$$
$$= -5.500$$

These dimensions have been inserted into Fig. 16.7(d).
(b) The corrected dimensions are as follows:

At ⊕:
$$X_c = -0.250$$
$$Z_c = 0.000$$

At A:
$$X_c = 0.750 - 0.250$$
$$= 0.500$$
$$Z_c = 0.000$$

At B:
$$X_c = 1.750$$
$$Z_c = -1.000 - 0.250$$
$$= -1.250$$

At C [Fig. 16.7(e)]:
$$\theta° = 90° - \frac{(180° - 50°)}{2}$$
$$= 25°$$

and
$$Z = 0.250 \tan 25°$$
$$= 0.117$$

Therefore,:
At C:
$$X_c = 1.750$$
$$Z_c = -2.250 - 0.133 = -2.383$$

At D [Fig. 16.7(f)]:
$$DB = 0.250$$

TABLE 16.3

N Address	G Code	X Axis	Z Axis	I Arc	K Arc	Spindle Speed (rev/min)	Feed (in./min)	M Function	Comment
N10						150	0.01	M04	Spindle on: CCW
N20	G01	0.5	-0.5						SP to A
N30			0.0						SP to A
N40	G02	1.75	-1.25	1.0	1.25				A to B
N50	G01		-2.383						B to C
N60		3.464	-3.866						C to D
N70	G02	4.0	-4.971	0.958	0.803				D to E
N80	G01		-5.750						E to F
N90	G00	4.3							Off work
N100		-0.25	0.0						To SP
N110								M02	(end of program)

Sec. 16.3 Radius–Angle Combinations

TABLE 16.4 PREPARATION COMMANDS: THE LATHE

Code[a]	Command
G00	Rapid traverse
G01	Linear interpolation: cut parallel to X or Z axes
G02	Circular interpolation: clockwise
G03	Circular interpolation: counterclockwise
G04	Dwell
G25	Subroutine
G27	Tool data
G33	Threads
G40	Cancel tool nose radius (TNR) compensation
G41	TNR compensation (cut above centerline)
G42	TNR compensation (cut below centerline)
G64	Radius cycle
G65	Angular cycle
G66	Contour cycle Z
G67	Contour cycle X
G68	Rough cycle Z
G69	Rough cycle X
G81	Turning cycle
G82	Facing
G83	Deep hole
G88	Grooving shaft
G89	Grooving face
G94	Feed (in./min)
G95	Feed (in./rev)

[a]Nonmodal functions affect the line in which they appear. Modal functions remain in effect until another function appears. The student should check the manufacturer's manual for verification.

$$CB = 0.250 \sin 40° = 0.161$$
$$DC = 0.250 \cos 40° = 0.147$$
$$X(\text{correction}) = CC' - CB = 0.250 - 0.161 = 0.089$$
$$Z(\text{correction}) = DA - DC = 0.250 - 0.147 = 0.103$$

TABLE 16.5 MISCELLANEOUS FUNCTIONS: THE LATHE

Function	Command
M00	Program stop
M01	Option stop
M02	End of program
M03	Spindle start forward (CW)
M04	Spindle start forward (CCW)
M05	Spindle stop
M06	Tool change
M08	Coolant on
M09	Coolant off
M16	Fully extends tailstock quill
M17	Fully retracts tailstock quill
M23	Tailstock unclamps
M24	Tailstock retracts
M30	End of tape
M36	Chuck open
M37	Chuck close
M38	Spindle stop

Therefore:
At D:

$$X_c = 3.553 - 0.089 = -3.464$$
$$Z_c = -3.763 - 0.103 = -3.866$$

At E:

$$X_c = 4.000$$
$$Z_c = -4.721 - 0.250 = -4.971$$

At F:

$$X_c = 4.000$$
$$Z_c = -5.500 - 0.250 = -5.750$$

These dimensions are inserted into Fig. 16.7(d).

(c) The program for Fig. 16.7(a) is shown in Table 16.3.

Summaries of commands for controlling the lathe are given in Tables 16.4 and 16.5.

QUESTIONS AND PROBLEMS

16.1. Describe the difference between radius compensation on the lathe and the milling machine. Illustrate using diagrams.

16.2. Discuss the four drawings shown in Fig. 16.1. You should be able to draw them from memory.

16.3. Describe the difference between the correction factor (k) and compensation (c).

16.4. Figure 16.8 represents a stepped cylinder. It is to be machines with a tool bit that has a TNR of $\frac{1}{16}$ in. Determine the coordinate X, Y coordinates at each junction point.

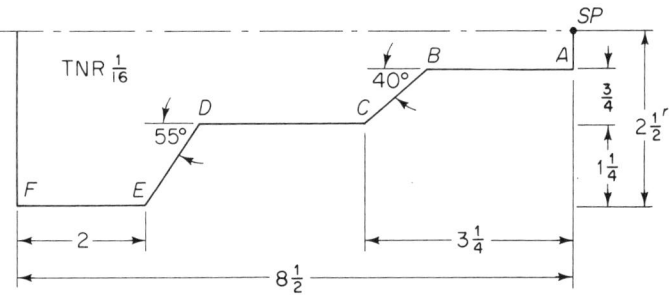

Figure 16.8

16.5. Program Fig. 16.9 using a TNR of $\frac{1}{8}$ in., a feed of 0.015 in./rev, and a spindle speed of 250 rev/min.

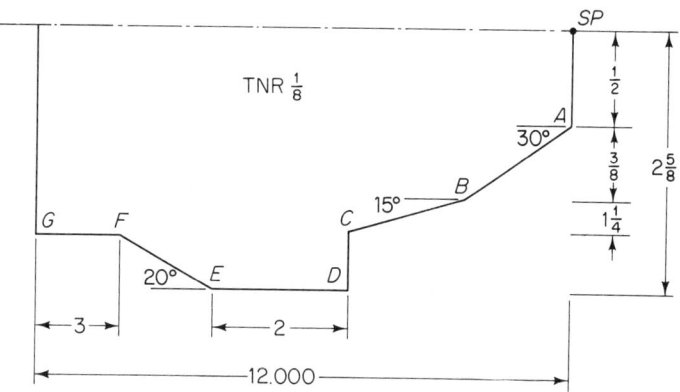

16.6. Program Fig. 16.10.
  (a) Calculate the junction points in the absolute system without TNR.
  (b) Calculate the corrections at each junction using a TNR of $\frac{3}{8}$ in.

276   The Lathe: Radius Compensation   Chap. 16

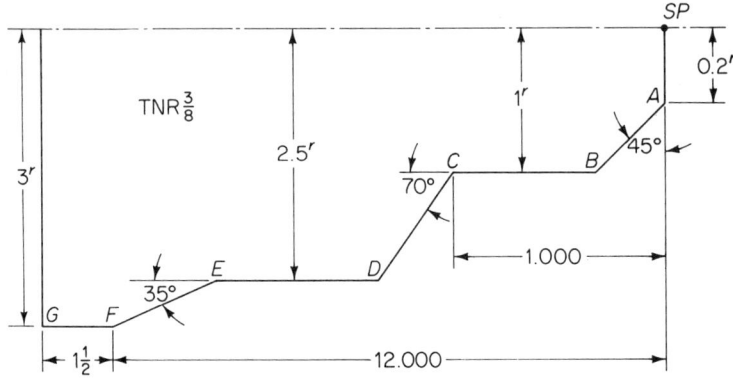

Figure 16.10

(c) Make a new drawing and insert the corrected dimensions at each junction.
(d) Write the program. Use a feed of 0.10 in./rev and a speed of 275 rev/min.

16.7. Repeat Prob. 16.6 with a TNR of $\frac{1}{4}$ in. Use a spindle speed of 0.030 in./rev and a spindle speed of 200 rev/min.

16.8. Repeat Prob. 16.6 for Fig. 16.11. Use a TNR of $\frac{1}{8}$ in., a feed of 0.015 in./rev, and a spindle speed of 300 rev/min.

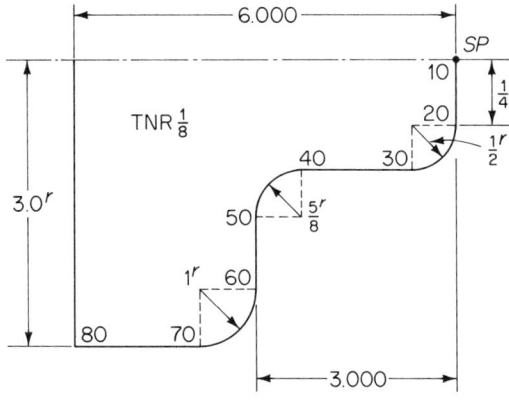

Figure 16.11

16.9. Repeat Prob. 16.6 for Fig. 16.12. Use a TNR of $\frac{1}{32}$ in., a feed of 0.010 in./rev, and a spindle speed of 250 rev/min.

Chap. 16 Questions and Problems   277

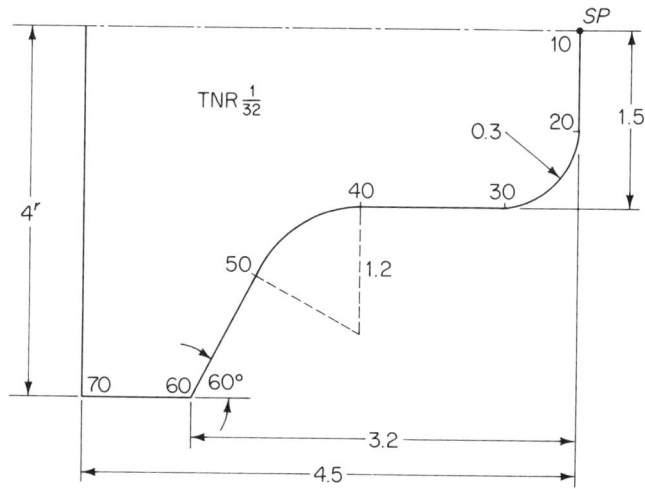

16.10. Repeat Prob. 16.6 for Fig. 16.13 using a TNR of $\frac{1}{32}$ in. Use feed of 0.030 in./rev, and a spindle speed of 150 rev/min.

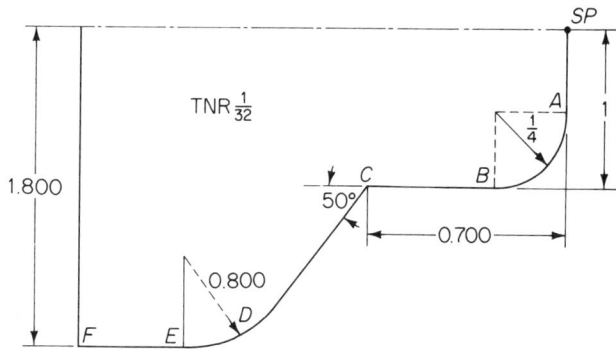

16.11. Repeat Prob. 16.6 for Fig. 16.14 using a TNR of $\frac{1}{32}$ in. Use a feed of 0.016 in./rev. and a spindle speed of 250 rev/min.
16.12. Calculate the compensation in Fig. 16.15. Use a TNR of $\frac{1}{4}$ in.
16.13. Calculate the compensation at each junction point and program Fig. 16.16. The TNR is $\frac{1}{4}$ in. The spindle speed is 400 rev/min and the feed is 20 in./rev.

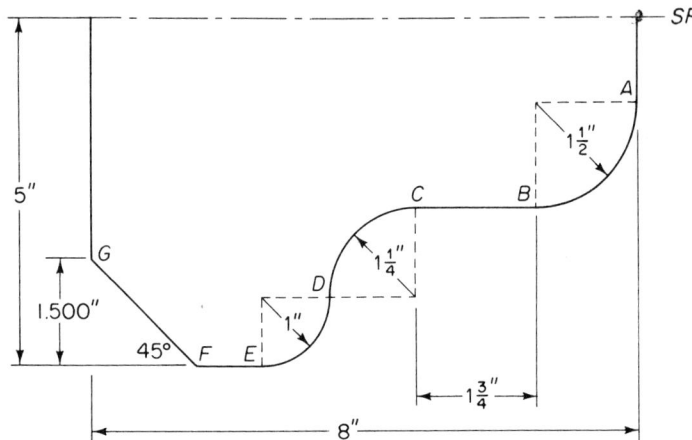

Figure 16.14

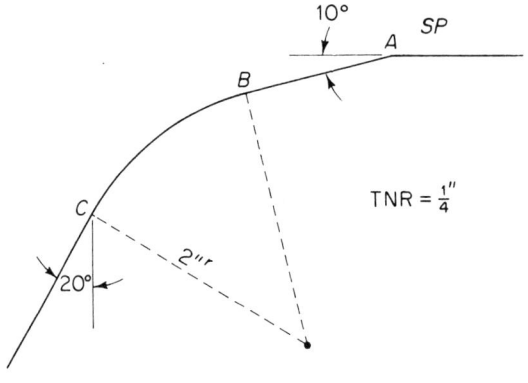

Figure 16.15

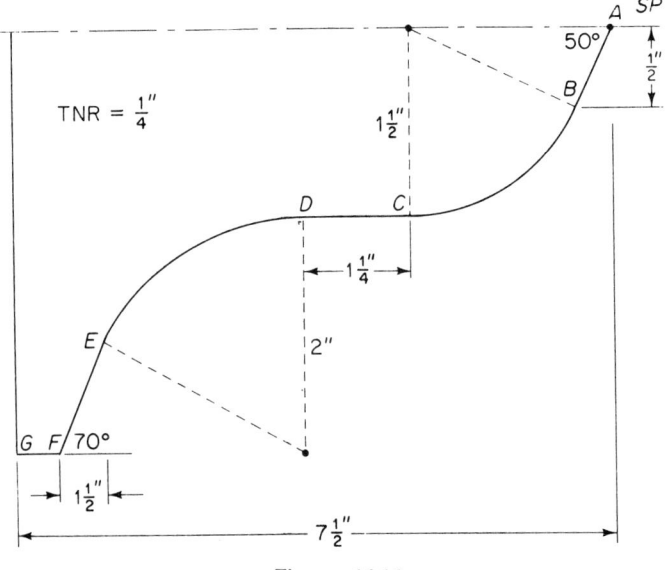

Figure 16.16

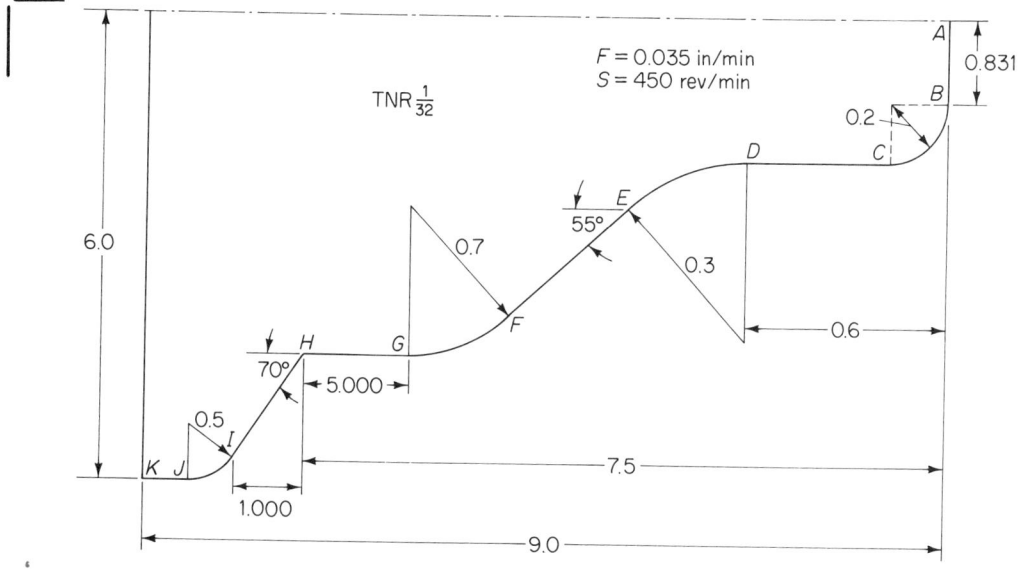

Figure 16.17

16.14. Program Fig. 16.17 using a TNR of $\frac{1}{32}$ in. and the absolute mode.
  (a) Calculate the junction points without compensation.
  (b) Calculate the corrections at each junction.
  (c) Apply these corrections to the X, Y values obtained in part (a).
  (d) Write the program.

16.15. Repeat Prob. 16.14 for Fig. 16.18.

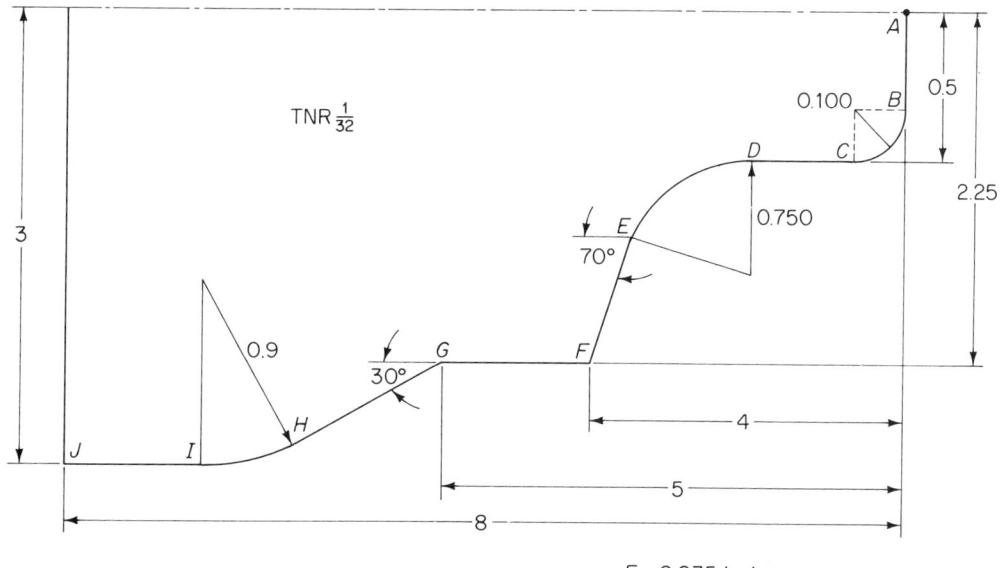

Figure 16.18

GLOSSARY

Absolute dimensions In a coordinate system of axes, each dimension is related directly to one start point (zero point); or to the origin of a coordinate axes system.

Address A letter, or series of letters or numbers, that establishes the location of a control command. For example: Y-5.2 indicates that the $-5,2$ is related to the Y-axis of a coordinate system. It may be related to the origin of a coordinate system; or to the last Y-value

Algorithm Defined rules, or processes, used to solve problems.

Alpha numeric code A code comprised of letters, numbers and specific characters.

Analog A system that represents axes positions.

ANSI American National Standards Institute

APT Automatic programmed tools.

ASCII American Standard Code for Information Interchange.

Automap Automatic Machine Programming.

Autospot Automatic Systems Positioning of Tools.

Auxiliary Functions Auxiliary functions to a program such as feeds, speeds, routines, selection of tools, etc.

Axis A reference line in a coordinate system.

Batch processing The use of a set of instructions before the next set is activated.

Binary code Symbol for two positions. 0 and 1 may mean on and off.

Binary coded decimal (BCD) A system that uses two numbers (0 and

Glossary

1) to represent numbers in the decimal numbering system. An example would be 0101.

Binary numbers Numbers that are written to base 2 and use 0 and 1. Thus
$$1101_2 = 1^3 1^2 0^1 1^0 = 1 \times 2^3 + 1 \times 2^2 + 0 \times 2^1 + 1 \times 2^0 =$$
$$(1 \times 8) + (1 \times 4) + (0 \times 2) + (1 \times 1) = 13$$
Therefore $1101_2 = 13$

Bit The absence, or presence, of a plus or minus charge on a tape. The smallest unit of data in a digital language.

Block A group of words that represent a complete operational instruction.

Byte A set of related bits of information.

CAD Computer Aided Design

CAM Computer Aided Manufacturing

Canned cycles A series of operations initiated by a single command.

Cartesian coordinates The system that defines a point in space relative to a zero point called the ordinate. It may be the X, Y, or Z distance taken from the intersection point of three axes at right angles to each other.

Character Letters, digits, or symbols used together and programmed into a computer to relay a specific message.

Circular interpolation A block, or blocks, of information that control radius arc generation.

Closed loop A system where the output command is fed back through the control computer to verify the input.

CNC Computer numerical control uses information fed into a computer to control the motion of a machine part, or tool.

Command A signal, or group of signals, initiating a step in the execution of a program.

Cursor A moveable pointer used to identify positions on a CRT.

Cutter compensation An adjustment to the programmed cutter path of a tool or cutter necessitated by tool wear, or cutter change.

Cutting speed The relative motion of a tool to the movement of the work in units of feet per minute.

Decimal numbers Numbers that use the base ten to define a magnitude. Thus
2456.5 may be written 2456.5_{10};
or
$$5 \times 10^{-1} + 6 \times 10^0 + 5 \times 10^1 + 4 \times 10^2 + 2 \times 10^3$$
$$= 2456.5$$

Digital The numbers 0 through 9.

DNC Direct Numerical Control. Data is fed directly into the machine from a large computer.

Do-Loop Similar to a canned cycle. It permits programs that repeat multiple operations. It is broader than a canned cycle in that any movement is stored in memory and called up by the beginning and end points of the stored program.

Dwell A delay in a program created by a command.

Editor A program that permits the creation of new files, or changing existing files.

End of Program This command indicates the completion of a program and causes the tape to rewind, coolant to shut off.

EOB Indicates the completion of a command, or series of commands.

Feed rate number The f-code that indicates a specific relationship between the movement of the tool and the work.

Feed rate override A command that allows for changing the programmed feed rate. It is usually manually directed.

Fixed canned cycles A pre-programmed series of repetitive operations.

Fixed cycles A pre-programmed command that instructs the machine to do a specific operation, or sequence of operations.

Floating zero The characteristic of a computer controlled machine to establish a zero point anywhere within the limits permitted by a machine tool.

Format The arranging of data in a formal manner.

fpm Feed in units of feet per minute.

FRN Feed rate number.

Hard copy A printed program.

Hardware The computer, a machine tool, that is hard wired and must be rewired if a change is needed. (See Software).

Initializing To reset the logic system to the beginning of a program.

Incremental dimension A dimension related to the preceding dimension as a reference point; ie (0,0).

ipm Inches per minute. Feed rate.

ipr Inches per revolution. Feed rate.

ipt Inches per tooth. Feed rate.

Interface The connection of a controller to a control system at a different location. This may be a larger computer.

Interpolation A series of points, between two desired end points, that produce a straight line, or curve.

Library A collection of frequently used programs, or routines.

Linear interpolation The distance along a straight line axis from one point to another.

Looping The process of jumping back to an earlier part of a program in order to repeat it.

Loop A sequence of repeated instructions used to produce an end result.

Glossary

Macros The process of changing a value of a program and thus producing a sub-program.

Manual data input (MDI) A mode that permits an operator to insert data into the control system manually.

Manual feed rate override A control that permits the operator to manually override the programmed feed rate.

Miscellaneous functions Controlled functions that are of an on-off auxiliary nature. One example may be coolant on-off. Another is spindle on clockwise (CW), or counterclockwise (CCW).

Modal A mode that remains in effect until changed.

Nesting A technique where a portion of a larger program is executed repetitively until a specific condition is encountered.

Offline Operations, or commands, that are not part of a program run on a numerically controlled machine. They may be performed by an auxiliary equipment.

Open loop A system that does not compare the input of a command with the output command.

Optional stop A program stop that requires an operators action in order to stop the program.

Parity May be odd or even. It requires all code commands in a taper to have either an odd or an even number of holes.

Part programmer One who writes a program to be input into a computer, either manually, or with a keyboard.

Pecking cycle A repetitive drilling operation program used for deep hole drilling where the operation requires clearing the chips before the drilling cycle operation is completed.

Point-to-point positioning The positioning, or control, of the movement of the tool, table, or spindle from one point to another.

Polar coordinates The location of a point, line, or plane as defined by a radius from a point and an angle from a reference point.

Postprocessor The unit that interprets instructions fed into a computer that converts the commands into movements of the tool, or work, as required by the program.

Preparatory function Commands that change the operations as required by the program.

Program A complete plan for the execution of a series of operations.

Program stop A miscellaneous function that stops the feed and spindle rotation at the completion of a portion of a program.

Rapid traverse The rapid positioning from one point to another.

Rectangular coordinates The straight line distances from a reference point as defined by the X, Y and Z directions from a zero point.

Retrofit Manually controlled machines that have been converted to numerical control.

RPM Spindle speed; revolutions per minute.

Sequence number A number in a program assigned to a block of information.

Setpoint The starting point of a program from which the first move is made.

sfpm Surface feet per minute. Cutting speed as referred to the linear movement of a spindle or table.

Software Programs that can be modified by changing instructions. (See hardware).

Spindle speed The rotation speed of a machine spindle. The units are revolutions per minute.

Storage A device that is capable of storing information so that it can be retrieved when needed.

Subroutines A programmed sequence of operations, stored in memory so that they may be called up as needed.

Tool function A program command that selects and calls up a tool.

Tool length offset (TLO) A distance between the tool and the work surface.

Tool nose radius compensation A correction that addresses the radius offset of a cutter, or toolbit.

Tool offset The compensation for tool length variations.

Word A set of digits that produce an instruction to the computer.

Zero offset This addresses the shift of the fixed zero position of a numerically controlled machine.

Index

A

Absolute mode:
  circular interpolation, 173–93
    multiquadrant circular
      interpolation, 180–91
  milling machine, 132–43
Absolute system, 92–93, 132–43
ADAPT word language, 85
Address, definition of, 72
American Standard Code for Information Interchange (ASCII) code, 8–9, 89
Angular compensation, 200–207
APT I, II, III word languages, 85
ASCII code, *See* American Standard Code for Information Interchange (ASCII) code
Automatic adaptive control (AAC), 12–13
AUTOPROMPT word language, 85
AUTOSPOT word language, 85

B

Back rake angles, 53
Begin code, 90
Binary-coded-decimal (BCD) system, 88
Blocks, definition of, 72
Boundary cut, 242
Bubble memory, 3
Built-up edges, 45–46

C

Canned cycles, 231–48
  circular pocket milling, 237–41
  cycles, 231
  multiple-row drilling, 235–37
  pocket milling, 241–44
  polar coordinates, 245–48
  surface milling, 231–35
Carbides, 50
Ceramic tool material, 50
Chip breakers, 55–57
  varieties of, 56
Circles, 15

Index

Circular contouring, incremental mode, 116–26
Circular interpolation, 207–12, 260–65
  absolute mode, 173–93
Circular pocket milling, canned cycles, 237–41
Closed-loop control systems, 5, 84–85
Codes:
  D and H codes, 80–81
  definition of, 72
  EIA, 8–9
  F code, 78–79
  G code, 74–75
  I, J, and K codes, 77–78
  incremental computer numerical control, 88–92
    begin code, 90
    types of, 88
    word address codes, 90
  M code, 80
  S code, 78
  T code, 79–80
  X, Y, and Z codes, 75–76
Command systems, 83–87
  incremental computer numerical control, 83–87
    closed-loop control systems, 84–85
    continuous-path control systems, 87
    open-loop control systems, 83–85
    semiclosed-loop control systems, 85
Computer mathematics, *See* Mathematics
Computer numerical control, 1–14
  automatic adaptive control (AAC), 12–13
  codes, 8–9
  computer-aided design (CAD)/computer-aided manufacturing (CAM), 10–12
  computer-integrated manufacturing (CIM) systems, 10–12
  direct numerical control (DNC), 9–10
  history, 1–3
Computer numerical control (CNC), 7
Continuous chip, 45
Continuous-path control systems, 87
Control centers, 64–71
  control panel, 64
  cycle start, 69
  emergency stop, 69
  feed hold, 69
  floating zero, 66–67
  manual control, 66
  manual data input (MDI), 64–66
  single block, 69
  tool length offset (TLO), 67–69
Conventional milling, compared to down milling, 59–60
Cosines, 22–23
Cratering, 46
Cutting speeds, 30–32
  and feed tables, 31–32
Cutting tools, 44–63
  drills/reamers, rake angle, 61
  milling cutters, 57–60
    down milling, 59–60
    drills/reamers, 60–61
    land, 57–58
    positive radial rake, 59
    relief angles, 59
  terminology, 51–57
    chip breakers, 55–57
    clearance, 53–54
    rake angles, 53–54
  theory of cutting, 44–48
  tool bit material, 48–51

Index

carbides, 50
ceramic tool material, 50
high-carbon steel, 49
high-speed steel, 49
industrial diamonds, 50–51
nonferrous cast tool steels, 49–50
Cycle start, 69

D

D code, 80–81
Direct numerical control (DNC), 3, 9–10
Discontinuous chip, 45
Down feed, 160
Down milling, compared to conventional milling, 59–60
Drill press:
   feed rate, 33–35
   horsepower requirements, 40–41
   time to machine, 33–35
Drills/reamers, 60–61
   rake angle, 61

E

Edge-cutting angles, 54
Electronic Industries Association (EIA), 88
   code, 8–9
Emergency stop, CNC machine tools, 69
End of block (EOB) carriage return, 89–90

F

F code, 78–79
Feed, 160–62
Feed command, 90
Feed rate, 32–37
   drill press, 33–35
   lathe, 33
   milling machine, 35–37
Finishing boundary, 242
Five/twelve/thirteen-degree triangles, 20–21
Flank wear, 46
Flexible manufacturing systems (FMS), 12
Floating zero, 66–67
Forty-five-degree triangles, 18–19
Free hold/free hold button, 69

G

G code, 74–75

H

Hard copy, 7
H code, 80–81
High-carbon steel, 49
High-speed steel, 49
Horsepower requirements, 37–41
   drill press, 40–41
   lathe, 37–39
   milling machine, 39–40

I

I code, 77–78
Incremental computer numerical control, 83–115
   codes, 88–92
      begin code, 90
      types of, 88
      word address codes, 90
   command systems, 83–87
      closed-loop control systems, 84–85
      continuous-path control systems, 87
      open-loop control systems, 83–85

Incremental computer numerical control (*cont.*)
  semiclosed-loop control systems, 85
    cutter position, 94–95
    X and Y incremental movements, 94–95
    linear contouring, 95–99
    program, 92–94
    Z movement, 99–108
Incremental mode, circular contouring, 116–26
Incremental systems, 92–93, 132
Industrial diamonds, 50–51

J

J code, 77–78

K

K code, 77–78

L

Land, 57–58
  milling cutters, 57–58
Lathe:
  feed rate, 33
  horsepower requirements, 37–39
  radius compensation, 252–75
    circular interpolation, 260–65
    linear contouring, 252–60
    radius-angle combinations, 265–74
  time to machine, 33
Law of cosines, 22–23
Law of sines, 21–22
Lead, tap, 159
Linear contouring, 95–99, 252–60
Linear interpolation, drill routines, 145–52

Lines, 73–74
  definition of, 72
Lubricating oils, cutting tools and, 48

M

M code, 80
Machine setup data (MSD), 69
Magnetic tapes, 7
Manual control, 66
Manual data input (MDI), 7, 64–66
Mathematics, 15–29
  circles/straight lines, 15
  cosines, law of, 22–23
  milling machine, 23–28
    ninety-degree tool movement, 23–24
    tool movements other than ninety degrees, 24–28
  right triangles, 16–18
  sines, law of, 21–22
  special relationships, 18–21
    five/twelve/thirteen-degree triangles, 20–21
    forty-five-degree triangles, 18–19
    thirty/sixty/ninety-degree triangles, 19–20
Milling cutters, 57–60
Milling machine:
  absolute mode, 132–43
  feed rate, 35–37
  horsepower requirements, 39–40
  radius compensation, 194–214
  time to machine, 35–37
  TNR compensation, 23–28
Modal G codes, 74
Multiple-row drilling, canned cycles, 235–37
Multiquadrant circular interpolation, 180–91

Index

N

Negative rake, 55, 56
Ninety-degree movements, 194–99
Nonferrous cast tool steels, 49–50
Nonmodal G codes, 74–75
Numerical control (NR), 2, 3–7

O

Oblique cutting, 48
Open-loop control systems, 5, 83–85
Orthogonal cutting, 48

P

Paper tape, 5, 7–8
Parsons, John, 1
Pocket milling, canned cycles, 241–44
Polar coordinates, 245–48
Positive rake, 55
Printout, 7
Programmable cycle files, 162–68
Programs, definition of, 72–73
Pythagorean triangle, 17–18

R

Radius-angle combinations, 265–74
Radius-angle compensation, 215–21
Radius-angle-radius compensation, 221–26
Radius compensation, 194–214
   angular compensation, 200–207
   circular interpolation, 207–12
   lathe, 252–75
   ninety-degree movements, 194–99
Rake angle, 61

drills/reamers, 61
side rake angles, 53
Read-only memory (ROM), 2
Reamers, 60–61
Relief angles:
   milling cutters, 59
   side relief angles, 53–54
Right triangles, 16–18
Rockwell hardness, tool materials, 50

S

S code, 78
Secondary clearance, 58
Side cutting-edge angles, 53
Side rake angles, 53
Side relief angles, 53–54
Sines, 21–22
Single block, 69
Single-point tools, 51
SNAP word language, 85
Spindle speed rate, 91
Spot drill, 145–52
Straight lines, 15
Surface milling, 231–35

T

Tapping, 158–62
   feed, 160–62
   lead, 159
   revolutions per minute, 159
T code, 79–80
Thirty/sixty/ninety-degree triangles, 19–20
Time to machine:
   drill press, 33–35
   lathe, 33
   milling machine, 35–37
Tool bit material, 48–51
   carbides, 50
   ceramic tool material, 50

Tool bit material (*cont.*)
   high-carbon steel, 49
   high-speed steel, 49
   industrial diamonds, 50–51
   nonferrous cast tool steels,
     49–50
Tool length offset (TLO), 67–69
Tool positioning:
   programmable cycle files,
     162–68
   tapping, 158–62
   and tool length offset, 157–58
Tungsten carbides, 50

W

Wilkenson, John, 1
Word address codes, 90

Work hardening, 45

X

X code, 75–76

Y

Y code, 75–76

Z

Z code, 75–76
Z movement, 99–108
Zero rake, 55